Social Network-Based Recommender Systems

Daniel Schall

Social Network-Based
Recommender Systems

 Springer

Daniel Schall
Siemens Corporate Technology
Wien, Austria

ISBN 978-3-319-37229-7   ISBN 978-3-319-22735-1 (eBook)
DOI 10.1007/978-3-319-22735-1

Springer Cham Heidelberg New York Dordrecht London
© Springer International Publishing Switzerland 2015
Softcover re-print of the Hardcover 1st edition 2015

Printed on acid-free paper

Springer International Publishing AG Switzerland is part of Springer Science+Business Media (www.
springer.com)

To my son Kilian

Preface

People increasingly use social networks to manage various aspects of their lives such as communication, collaboration, and information sharing. A user's network of friends may offer a wide range of important benefits such as receiving online help and support and the ability to exploit professional opportunities. One of the most profound properties of social networks is their dynamic nature governed by people constantly joining and leaving the social networks. The circle of friends may frequently change when people establish friendship through social links or when their interest in a social relationship ends and the link is removed.

This book introduces novel techniques and algorithms for social network-based recommender systems. Here, concepts such as link prediction using graph patterns, following recommendation based on user authority, strategic partner selection in collaborative systems, and network formation based on "social brokers" are presented. In this book, well-established graph models such as the notion of hubs and authorities provide the basis for authority-based recommendation and are systematically extended towards a unified Hyperlink Induced Topic Search (HITS) and personalized PageRank model. Detailed experiments using various real-world datasets and systematic evaluation of recommendation results proof the applicability of the presented concepts.

Vienna, Austria   Daniel Schall
June 2015

Acknowledgements

This book provides a detailed summary and new viewpoints on the author's research in the field of social network analysis and link formation techniques. Daniel Schall received his Ph.D. degree in computer science from the Vienna University of Technology in 2009.

He started his research career at Siemens Corporate Research in Princeton, NJ, USA, in 2003, where he was employed as a technical associate. In 2006, he began his doctoral studies at the Vienna University of Technology. At the Vienna University of Technology, he was involved in a number of European funded FP6 and FP7 projects both as a project manager and key researcher. Daniel was a principal investigator of crowdsourcing and social computing activities at the Distributed Systems Group. The main results of his research in the context of crowdsourcing and Human Provided Services were published in the book *Service-Oriented Crowdsourcing: Architecture, Protocols and Algorithms*. Also, he published more than 40 scientific papers at top-ranked international conferences and more than 20 scientific journal papers in highly ranked journals and renowned magazines including the Journal of Informetrics, Decision Support Systems, Data and Knowledge Engineering, Information Systems, IEEE Transactions on Services Computing, IEEE Computer, IEEE Internet Computing, and Social Network Analysis and Mining.

Dr. Daniel Schall is currently employed as a senior key expert research scientist at Siemens Corporate Technology in Vienna, Austria.

Contents

Chapter 1
Overview Social Recommender Systems

Abstract This chapter gives an introduction to social network-based recommender systems. The main recommendation techniques as presented in this book including link prediction, follow recommendation, partner recommendation using reputation evaluation, and social broker recommendation are highlighted.

1.1 Recommendations in Social Networks

In recent years considerable attention has been devoted to the analysis of social networks structures. Social networks typically consist of nodes representing people or other entities embedded in a social context and edges representing interaction, collaboration, or some other form of linkage between entities. Examples of social networks include personal social networks such as Facebook, Twitter, Google Plus or professional social networks such as LinkedIn. Other social network based systems are, for example, the set of organizations collaborating in the context of research projects or organizations forming professional virtual communities. The availability of large, detailed social network datasets has stimulated extensive research of their basic properties. Social networks are highly dynamic and grow and change quickly over time through the addition of new edges or removal of existing edges.

Link formation techniques support the discovery and establishment of social relations in social network based systems. The applications include:

- *Establishment of New Social Relations.* The emergence of new personal relations is actively facilitated through link prediction techniques.
- *Supporting the Formation of Expert Communities.* Following recommendations are based on community expertise and increase the cohesiveness of online communities.
- *Strategic Formation of Teams.* The automatic discovery of community reputation and structural holes helps in establishing competitive compositions of research organizations.
- *Bridging Online Communities.* Fragmented communities are connected through social brokers who act as intermediaries between communities and strengthen information exchange.

© Springer International Publishing Switzerland 2015   1
D. Schall, *Social Network-Based Recommender Systems*,
DOI 10.1007/978-3-319-22735-1_1

The next section provides an introduction to main link formation techniques in social network based systems as detailed in this book.

1.2   Recommendation Techniques

Figure 1.1 gives an overview of the presented recommendation techniques. The techniques are structured into *peer* based recommendations (upper row in Fig. 1.1) and *group* based recommendations (lower row in Fig. 1.1). Each recommendation technique will be introduced in the following sections.

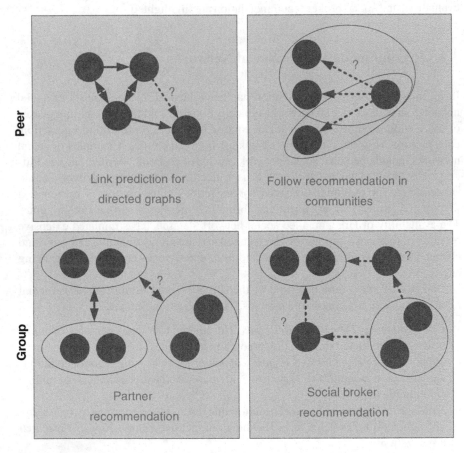

Fig. 1.1 Recommendation techniques overview

1.2.1 Link Prediction

In today's online social networks it becomes essential to help newcomers as well as existing community members to find new social contacts. In scientific literature this recommendation task is known as link prediction [6]. Link prediction has important practical applications in social network platforms. It allows social network platform providers to recommend friends to their users. Another application is to infer missing links in partially observed networks.

The meaning of a recommendation varies depending on the concrete social network platform. In platforms such as Facebook a link between two people is established if both persons agree to have a friendship relation. The resulting network is thus *undirected* because both persons share mutual friendship. Another example of a social network is Twitter. In Twitter a link between two persons is established if a user is interested in news updates of another user. The link is thus directed because there is no mutual agreement needed to establish a link. The resulting network is *directed* and is also called follower network. User can follow an arbitrary number of other users to receive news or activity updates. The shortcoming of many of the existing link prediction methods is that they mostly focus on undirected graphs only [7, 14].

1.2.2 Follow Recommendation

Open source development allows a large number of people to reuse and contribute source code to the community. Social networking features open opportunities for information discovery, social collaborations, and improved recommendations of potential collaborators. Online community and development platforms rely on social network features to increase awareness and attention among community members for improved collaborations.

In networks such as LinkedIn or Facebook friendship is represented as reciprocated links in an undirected graph. Services such as Twitter and recently GitHub are based on a directed network approach. A directed network approach allows users to follow other users based on their interest without requiring them to reciprocate the relationship. In traditional social networks, some users may be followed by many people without following many peers themselves ("stars" or "celebrities"). Is this also the case for online social collaboration networks such as GitHub? People in GitHub are mostly followed because they work on interesting projects. Thus, this difference between conventional social networks and online social collaboration networks requires a novel "who to follow" recommendation approach [13].

1.2.3 Partner Recommendation

Scientific collaborations commonly take place in a global and competitive environment. Coalitions and project consortia are formed among universities, companies and research institutes to apply for research grants and to perform jointly collaborative projects. In such a competitive environment, individual institutes may be strategic partners or competitors. Measures to determine partner importance have practical applications such as comparison and rating of competitors, reputation evaluation or performance evaluation of companies and institutes [10].

However, the success of research and innovation is based on the right balance between cooperation and competition. Hence, formation of coalitions and consortia is influenced by partner reputation, institutional constraints, and mechanism of self-organization. Scientific collaboration can be analyzed at the level of researchers through co-authorship and citation networks or at the level of organizations or research institutions [5]. The former has been widely studied by existing research while the latter lacks a principled approach for selecting and aggregating ranking criteria that may be influenced by context [11].

1.2.4 Broker Recommendation

The rapid advancement of ICT-enabled infrastructure has fundamentally changed how businesses and companies operate. Global markets and the requirement for rapid innovation demand for alliances between individual companies. A virtual organization can be defined as follows [4]: *an inter-organizational virtual organization is a temporary network organization, consisting of independent enterprises (organizations, companies, institutions, or specialized individuals) that come together swiftly to exploit an apparent market opportunity. As such, virtual organizations act in all appearances as a single organizational unit.*

Principles found in social network theory are promising candidate techniques to assist in the formation process and to support flexible and evolving interaction patterns in cross-organizational environments. In social networks, relations and interactions typically emerge freely and independently without restricted paths and boundaries. Research in social sciences has shown that the resulting social network structures allow for relatively short paths of information propagation [16]. While this is true for autonomously forming social networks, the boundaries of collaborative networks are typically restricted due to organizational units and fragmented areas of expertise. This demands for novel formation patterns such as brokers [1, 12].

1.3   Research Datasets

Most experiments performed in this research are based on real datasets. The main datasets used in this research are all publicly available and include:

- A *Twitter* dataset has been obtained from [15] and has been used to perform various experiments in the context of link prediction.
- A *Google Plus* dataset has been obtained from [15] and has also been used to perform link prediction.
- A *GitHub* dataset has been obtained from [2] and event information from [3]. The resulting dataset has been used to perform link prediction, and follow recommendations.
- A statistical report of research activities in the European's ICT program has been obtained from [8] and used to perform experiments in the context of strategic partner selection and social broker discovery.

The rich set of data collections provides a solid basis for performing comprehensive experiments to derive insights and key findings. In the following, an outline of the book is given.

1.4   Book Outline

This book is organized as follows. In Chap. 2 we introduce link prediction methods and metrics for directed graphs. We compare well known similarity metrics and their suitability for link prediction in directed social networks. Chapter 3 introduces our "who to follow" recommendation model. Link analysis techniques such as PageRank and HITS provide the basis for a novel "who to follow" recommendation model. In Chap. 4 we present a novel approach for measuring and combing various criteria for partner importance evaluation. The presented approach is cost sensitive, aware of temporal and context-based partner authority, and takes structural information with regards to structural holes into account. Chapter 5 focuses on the notion of brokers who act as intermediaries between segregated communities. We introduce a broker discovery and ranking approach utilizing a link-based broker importance model. Finally, the book is concluded in Chap. 6.

The work presented in this book is based on the author's research performed over the last years. Prior work of the author includes research in crowdsourcing, online communities and social network analysis (see [9]). This work seamlessly goes into similar fields of research and expands more deeply in social networks and link formation techniques in social network based systems.

Some material has been adapted for this book based on the author's journal publications [11–14]. All material has been revisited, additional experiments have been performed, and further enhancements in the concepts and implementation have been done. Among others, new concepts include machine learning extensions in the link prediction framework, the notion of an app and service marketplace in hybrid compute environments, and corporate policies in the context of partner discovery and selection.

References

1. R. S. Burt.   Structural Holes and Good Ideas.   *The American Journal of Sociology*, 110(2):349–399, 2004.
2. GitHub. Online: www.github.com (last access June 2015).
3. I. Grigorik. Online: www.githubarchive.org (last access June 2015).
4. E. C. Kasper-Fuehrera and N. M. Ashkanasy. Communicating trustworthiness and building trust in interorganizational virtual organizations. *Journal of Management*, 27(3):235–254, June 2001.
5. N. Lavrac, P. Ljubic, T. Urbani, G. Papa, M. Jermol, and S. Bollhalter. Trust modeling for networked organizations using reputation and collaboration estimates. *IEEE Transactions on Systems, Man, and Cybernetics, Part C*, 37(3):429–439, 2007.
6. D. Liben-Nowell and J. Kleinberg. The link prediction problem for social networks. In *Proceedings of the twelfth international conference on Information and knowledge management*, CIKM '03, pages 556–559, New York, NY, USA, 2003. ACM.
7. L. Lu and T. Zhou. Link prediction in complex networks: A survey. *Physica A: Statistical Mechanics and its Applications*, 390(6):1150–1170, 2011.
8. F. Munisteri. ICT statistical report for annual monitoring 2011. http://ec.europa.eu/digital-agenda/sites/digital-agenda/files/stream_2012_0.pdf, Feb. 2012.
9. D. Schall. *Service Oriented Crowdsourcing: Architecture, Protocols and Algorithms*. Springer Briefs in Computer Science. Springer New York, New York, NY, USA, 2012.
10. D. Schall. Measuring contextual partner importance in scientific collaboration networks. *J. Informetrics*, 7(3):730–736, 2013.
11. D. Schall. A multi-criteria ranking framework for partner selection in scientific collaboration environments. *Decision Support Systems*, 2013.
12. D. Schall. Socially-based brokerage and composition in virtual communities. *IJNVO*, 12(3):251–281, 2013.
13. D. Schall. Who to follow recommendation in large-scale online development communities. *Information and Software Technology*, 2013.
14. D. Schall. Link prediction in directed social networks. *Social Network Analysis and Mining*, 2014.
15. Stanford. Online: http://snap.stanford.edu/data/index.html (last access June 2015).
16. D. J. Watts and S. H. Strogatz. Collective dynamics of 'small-world' networks. *Nature*, 393(6684):440–442, June 1998.

Chapter 2
Link Prediction for Directed Graphs

Abstract In this chapter we introduce link prediction methods and metrics for directed graphs. We compare well known similarity metrics and their suitability for link prediction in directed social networks. We advance existing techniques and propose mining of subgraph patterns that are used to predict links in networks such as GitHub, GooglePlus, and Twitter. Our results show that the proposed metrics and techniques yield more accurate predictions when compared with metrics not accounting for the directed nature of the underlying networks.

2.1 Friendship Recommendation

Social networks have become ubiquitous in our everyday activities. People use social networks to communicate, collaborate, and share information. One of the most profound properties of social networks is their dynamic nature. People join and leave social networks. Also, the circle of friends may frequently change when people establish friendship through social links or when their interest in a social relationship ends and the link is removed. Due to the large number of users being part of today's online communities, it becomes increasingly cumbersome to find new contacts and friends. Many social network platform providers assist their users in establishing new social relations by making recommendations. The meaning of a recommendation varies depending on the concrete social network platform. In platforms such as Facebook [11] a link between two people is established if both persons agree to have a friendship relation. The resulting network is thus *undirected* because both persons share mutual friendship. Another example of a social network is Twitter [43]. In Twitter a link between two persons is established if a user is interested in news updates of another user. The link is thus directed because there is no mutual agreement needed to establish a link. The resulting network is *directed* and is also called follower network. User can follow an arbitrary number of other users to receive news or activity updates. The "follow" feature is in widespread use in social networking services such as Twitter [43], Facebook [11], or Google-Plus [14]. In Facebook, the users can follow (or unfollow) news updates of their friends. The social (undirected) link between friends is maintained irrespective of the follow relationship. Recently collaborative online platforms such as GitHub [12]

© Springer International Publishing Switzerland 2015   7
D. Schall, *Social Network-Based Recommender Systems*,
DOI 10.1007/978-3-319-22735-1_2

offer also social network features (e.g., following). GitHub is an online "coding" community. Users contribute code and share repositories with the community. The "follow" feature in GitHub allows users to keep track of updates regarding various software development activities such as coding or bug-fixing.

All of the before mentioned platforms have a large number of users and benefit from "follow" recommendations. Such recommendations can be formulated as a link prediction task. Link prediction in a directed follower network has the purpose to give recommendations who a given user should follow. Despite of some existing literature in the area of link prediction on Twitter (for example, see [8, 33]), or prediction of positive/negative edges [23], there is relatively little work on link prediction in directed social networks. As reported in a recent survey [26], the existing studies on link prediction overwhelmingly focus on undirected networks. This work specifically addresses the link prediction problem in directed social networks.

The essential approach we follow in this work is to measure the similarity between a pair of nodes using *structural* information. We assume that a social network is modeled as a directed graph $G(V,E)$ where vertices V in the graph depict people and edges E between vertices relations between people. Structural information is, for example, the number of common neighbors between two nodes. Generally speaking, similarity of nodes can be measured as the number of common features. The goal of link prediction is to predict whether a link between two users will be established or if a link in a partially observed network is missing. The latter case is a common problem when the social network is obtained through crawling.

Here we provide the following key contributions:

- We propose link prediction techniques using graph patterns (*triads*). Predictions are then given based on the probability that a given type of triad pattern will be closed.
- Here we introduce a metric called *Triadic Closeness*. The application of the metric is discussed and evaluated.
- We have designed and implemented a link prediction framework. The main elements of the framework are discussed.
- We present an analysis of our link prediction techniques. We use three different social networks including GitHub, GooglePlus and Twitter to validate the proposed approach.

This chapter is structured as follows: Sect. 2.2 discusses related work in the context of social networks and link prediction. Standard similarity metrics are introduced which will be compared against our proposed approach. In Sect. 2.3 the link prediction framework is introduced followed by a discussion on our prediction approach Sect. 2.4. The data collection is introduced in Sect. 2.5. The results and experiments are discussed in Sect. 2.6. The conclusion with outlook to further work is given in Sect. 2.7.

2.2 Background in Link Prediction

We discuss related work by (1) highlighting literature and related approaches with respect to link prediction and node similarity indices and (2) local structures and patterns in social networks. Similarity indices are structured in *local*, *global*, and *semi-local* indices.

- **Local Similarity Indices.** A wide range of similarity metrics exist that can be used to predict links based on "local" information [1, 2, 10, 22, 31, 32, 34, 46]. Local information is typically obtained by comparing degree of overlap of two individual friendship networks. Liben-Nowell et al. [24] systematically compared a number of local similarity indices in many real networks. These metrics focus on undirected graphs without considering directed relations. The advantage of local indices is that they can be computed for large-scale networks and do not require a huge amount of computational resources.
- **Global Similarity Indices.** Global metrics take the properties of the whole social network into account. The Katz index [20] is based on the ensemble of all paths between two nodes in the network. The index is computed as the sum over the collection of all paths and is exponentially damped by length to give the shorter paths more weight. Another class of metrics are random walk techniques. Well-known algorithms such as PageRank [30] can be used to compute global importance metrics. The prediction is than based on the node's PageRank importance score. PageRank can be personalized to perform ranking with respect to a certain "contexts" or topics [36]. A direct application of personalized PageRank are Supervised Random Walks (SRW). In [5], SRW were proposed to recommend links in networks such as Facebook. SimRank [18] is based on the idea that two nodes are similar if they are related to similar nodes.

  Global similarity indices naturally require information regarding the whole topology of the social network. Indeed, this information may not be available due to, for example, partially observed networks or in cases where the platform is decentralized. Another important aspect is performance and resource consumption. The calculation of global indices may be very time-consuming and for large-scale networks the computation may not be feasible.
- **Semi-Local Indices.** Instead of taking the whole topology into account, semi-local indices omit information that makes little contribution to improve the prediction algorithm's accuracy. The Local Path Index [28, 46], for example, provides a trade-off between computational complexity and accuracy. Local Random Walks [5, 25] follow a similar idea by omitting information from very distant neighbors in the network.

  Further methods for link prediction include hierarchical models [9], stochastic block models [3, 16, 45], probabilistic models [26], and methods considering positive/negative links [42] (see [26] for details on models and methods).
- **Patterns, Triads and Motifs.** The approach as proposed in this work takes the directed nature of follower networks into account. *Triadic closure* in social networks is the hypothesis that the formation of an edge between u and v is strongly dependent upon the degree of overlap of u's and v's individual

friendship networks (for example, see [39, 44]). However, an important theory in this context is the "strength of weak ties" [15] stressing the cohesive power of seemingly less important ties. Holland and Leinhardt [17] developed many important theories about social relations and how to detect structure in directed networks. As a more general conceptual framework, network motifs [4, 29] represent elementary elements in complex networks. Complex networks include social, technological, or biological networks. We build upon these ideas and propose link prediction considering triad patterns in social networks.

Finally, previously we stressed the importance of social networks and formations of social groups (teams) in the context of collaborative environments and novel crowdsourcing environments [35, 37, 38]. We foresee important applications of link prediction in these areas.

2.3 Software Framework

One of the goals of the present work has been to design a modular and reusable framework for link prediction in social networks. The framework must be able to handle different social networks that can range from a few thousands to millions of nodes in the social graph. This section gives an overview of the link prediction framework architecture and a description of the evaluation methodology. The framework is extensible with regards to prediction metrics and algorithms.

The main elements and layers are depicted by Fig. 2.1. The overall framework is segmented into a multilayer architecture consisting of three layers: (1) Data Layer, (2) Prediction Layer, and (3) Presentation Layer. Each layer has distinct responsibilities and each block within a given layer provides a well-defined set of interfaces. The central goal of the system is to provide an extensible framework that can be enhanced with new metrics and prediction methods. In addition, it should be easy to add new social networks (datasets) and to compare the results of different experiments and algorithms. The architectural layers are described in the following.

Data Layer The bottom-most layer is concerned with low-level data handling and persistence management. With regards to the basic link prediction task, the Data Layer passes an instance of a social network graph to the upper layer. Our framework has the ability to access data from (a) the *Flat File Store* and (b) the *Database*. The Flat File Store is used for simple social network data files that are small to medium in their size (e.g., 10^3–10^5 nodes) and is read-only. The *Parser* reads and interprets files stored in the Flat File Store. The Parser performs the pre-stage processing for the *Graph Mapper*, which creates the social network graph including node and edge attributes (if available). The Database is able to manage large social networks (large networks may consist of up to 5×10^7 nodes[1]) that may

---

[1]The upper bound for which the Data Layer has been tested was a network consisting of approximately 5×10^7 nodes and 1.5×10^9 edges.

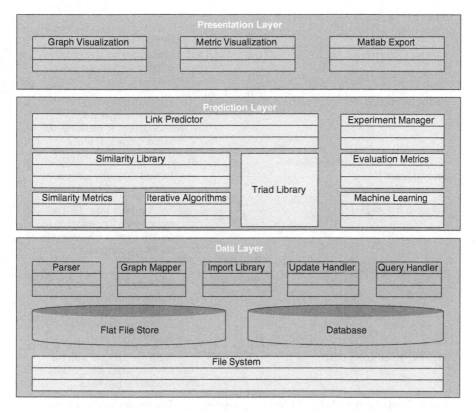

Fig. 2.1 Link prediction framework overview

also include details regarding user profiles and user activity. The *Query Handler* interfaces with the database to retrieve the social networks graph structure, user profile details, other social network related information and also information related to patterns and experiments. The *Update Handler* is responsible for persistence management and writes data to the database. The *Import Library* allows external social networks and networks stored in the Flat File Store to be migrated to the Database. Migration from the Flat File Store is needed if performance is insufficient (read operations) or if the management of social network (meta)data through the Flat File Store becomes impractical. Another source of information are external APIs of social network or community platforms. GitHub, for example, provides an API [13] to retrieve the follower network. The Import Library provides a rate-aware API invocation scheduler to retrieve large social networks respecting the social network providers' API policies (e.g., number of invocations per hour).

Prediction Layer The middle layer groups the logic for metric calculation, pattern mining (triad detection), prediction, and prediction result evaluation. The *Similarity Library* is responsible for calculating various similarity indices including local

Table 2.1 Similarity-based metrics

Metric	Definition	Description
Common Neighbors (CN)	$\lvert \Gamma(u) \cap \Gamma(v) \rvert$	Intersection set size of joint neighbors between nodes u and v
Salton Index (SA)	$\dfrac{\lvert \Gamma(u) \cap \Gamma(v) \rvert}{\sqrt{k_u \times k_v}}$	The degree of u and v is depicted by k_u and k_v respectively. In literature, the Salton index [34] is also called the cosine similarity
Jaccard Index (JA)	$\dfrac{\lvert \Gamma(u) \cap \Gamma(v) \rvert}{\lvert \Gamma(u) \cup \Gamma(v) \rvert}$	Jaccard similarity index with $\Gamma(u) \neq \emptyset$ and $\Gamma(v) \neq \emptyset$
Sørensen Index (SO)	$\dfrac{2\lvert \Gamma(u) \cap \Gamma(v) \rvert}{k_u + k_v}$	The Sørensen index [40] is mainly used for ecological data. The index is identical to the Dice's coefficient
Hub Promoted Index (HP)	$\dfrac{\lvert \Gamma(u) \cap \Gamma(v) \rvert}{\min(k_u, k_v)}$	The links adjacent to hubs are likely to be assigned higher scores since the denominator $\min(k_u, k_v)$ is determined by the lower degree [31]
Hub Depressed Index (HD)	$\dfrac{\lvert \Gamma(u) \cap \Gamma(v) \rvert}{\max(k_u, k_v)}$	Similar to HPI but with the opposite effect with regards to adjacent hub links
Leicht-Holme-Newman Index (LHN)	$\dfrac{\lvert \Gamma(u) \cap \Gamma(v) \rvert}{k_u \times k_v}$	The denominator $k_u \times k_v$ is proportional to the expected number of common neighbors [22]
Adamic-Adar Index (AA)	$\displaystyle\sum_{z \in \Gamma(u) \cap \Gamma(v)} \frac{1}{\log(k_z)}$	The index assigns the less-connected neighbors more weight than CN [1]
Resource Allocation Index (RA)	$\displaystyle\sum_{z \in \Gamma(u) \cap \Gamma(v)} \frac{1}{k_z}$	Similar to AA, RA depresses the contribution of the high-degree common neighbors [46]

similarity indices (see Table 2.1), global similarity indices, and semi-local indices. The system block *Similarity Metrics* computes local similarity indices (including TC) and semi-local indices including Local Path Index (see [26]) and the Shortest Path Index (i.e., the average Dijkstra Shortest Path Index between, say, u and v's neighbors $\Gamma(v)$ to measure similarity between u and v). *Iterative Algorithms* have been designed to calculate metrics such as SimRank [18] and PageRank variants such as personalized PageRank [19, 30, 36]. Note, the evaluation of global and semi-local indices is not within the scope of this work. Here we focus on local indices in conjunction with triad patterns. The relationship between (local) triad patterns and global or semi-local indices is a whole new subject of investigation itself. Next to the Similarity Library, the *Triad Library* provides the capabilities for triad pattern detection and caching. Clearly, scanning the entire social network graph for triad patterns is a time consuming task with complexity $\mathscr{O}(m)$ where m is the number of edges in the graph (see [6] for related triad detection algorithms).

Thus, triad pattern mining is usually performed once for a given social network and subsequent metric calculations use the precomputed triad frequencies.

The *Link Predictor* is the main component that performs the prediction task. This can be done to predict future links, which do not yet exist, or predict links, which are "missing" (unobserved). Here we focus on the latter case where we assume that certain links are missing between pairs of nodes. To test the metrics' accuracy, we divide the set of edges E randomly into the prediction (or training) set E^P and the validation set E^V. The prediction algorithm basis its calculation upon the prediction graph $G^P(V, E^P)$ whereas the accuracy of the prediction results is determined by inspecting the missing (randomly removed) edges in E^V. Note, no information from the set E^V is allowed to be used for prediction so that $E^P \cup E^V = E$ and $E^P \cap E^V = \emptyset$. Furthermore, to speed up computation of prediction results, the Link Predictor can perform node sampling to calculate predictions for a subset of node pairs instead of calculating predictions for all node pairs in the entire graph. For that purpose the predictor samples a set of random nodes U^P with $k > 0$ and divides the set into two subsets U^R and U^T with $U^R \cup U^T = U^P$, $U^R \cap U^T = \emptyset$, and $U^P \subset U$. The set U^P contains all nodes that are used for link prediction, the set U^R contains the root nodes (source vertices of predicted links) and U^T contains the target nodes (target vertices of predicted links). The set E^V contains only directed links whose source vertex is in U^R and whose target vertex is in U^T. For one given experiment the same node set U^P and edge set E^V is used to be able to compare the results of different metrics among each other.

The basic steps of the Link Predictor are straightforward. Algorithm 1 shows the steps:

Algorithm 1 Link prediction algorithm

1: **input:** $G(U, E^P), U^R, U^T, E^V$
2: **for each** User $u \in U^R$ **do**
3:     **for each** User $v \in U^T$ **do**
4:         // True Answer
5:         $answer \leftarrow$ HasEdge(u, v, E^V)
6:         // Calculate Similarity Scores
7:         **for each** Metric $m \in M$ **do**
8:             $s_{uv} \leftarrow$ CalculateScore(u, v, m, G)
9:             // Save Result to Experiment Database
10:             SaveResult$(u, v, m, s_{uv}, answer)$
11:         **end for**
12:     **end for**
13: **end for**

The predictor loops through U^R and U^T and calculates similarity scores s_{uv} using each metric $m \in M$ provided by the Similarity Library. M can be configured dynamically by enabling/disabling the desired metrics to be used in each experiment. The prediction result s_{uv} and the actual true answer $\{0, 1\}$ (*HasEdge* is a binary classifier that determines whether or not the set E^V contains a directed edge between u and v) are saved in the experiment database.

The *Machine Learning* component provides an additional approach to link prediction by learning ensembles of decision trees. This technique is called *random forest*[2] for classifying if there will be link between two nodes from an input vector of node features (e.g., number of common friends, age, interests, etc.). A detailed discussion on random forest based link prediction is not provided in this book.

To compare the results of different similarity algorithms, the *Evaluation Metrics* component provides standardized comparison methods. In particular, we compare results using the *Receiver Operating Characteristic* (ROC) curve. ROC curves are commonly used in the machine learning community for the link prediction task [2, 9]. ROC curves are created by plotting the true positive rate over the false positive rate. The *area under the ROC curve* (AUC) [7] can be interpreted as the probability that a randomly chosen missing link (i.e., a link in E^V) is given a higher score than a randomly chosen non-existing link [26].

Higher AUC values, which are in the range [0,1], indicate better prediction performance. Another common metric to measure a prediction algorithm's accuracy is *HitRatio* (or recall). Generally, HitRatio is defined as the ratio of selected relevant items to the number of relevant items. For example, the HitRatio is typically measured at a threshold HitRatio@n where n is the number of selected items. In this work, we focus on both ROC curves and AUC as well as HitRatio for experiment evaluation and metric comparison. The *Experiment Manager* saves and retrieves experiments results from the Database and computes aggregates of results.

Presentation Layer The frontend of the prediction framework is a presentation layer that has visualization and export capabilities. The *Graph Visualization* allows to view typical graph properties by mapping node/edge features into a visual representation. To do so, the correspondence between discrete or continuous values and visual properties (color, node size, etc.) needs to be established. The *Metric Visualization* is the most important tool for evaluating the results of the similarity algorithms and link predictor. ROC curves help to identify which methods and parameter settings are best suited for a given type of social network. The various network idiosyncrasies such as average degree $\langle k \rangle$ may demand for metric tuning. A detailed discussion regarding metric accuracy will follow in Sect. 2.6. The Metric Visualization helps to understand the accuracy and suitability of different metrics. The *Matlab Export* allows to export experiment results to a Matlab compatible format to utilize various Matlab toolboxes.

---

[2]Web page: https://www.stat.berkeley.edu/~breiman/RandomForests/cc_home.htm.

2.4 Predicting Friendship

2.4.1 Similarity-Based Metrics

The focus of this work are local similarity indices and their extension towards link prediction using graph patterns. Table 2.1 lists a set of well-known similarity metrics. The shared feature of the metrics is that computation of similarity is based on the set of joint neighbors. These metrics provide the basis for the definition of our *Triadic Closeness* (TC) similarity metric. The definition of the TC metric will be provided in the following. Furthermore, the metrics in Table 2.1 will be used in a comparative study to test the effectiveness of the proposed technique. Table 2.1 provides a mathematical definition along with a brief description of the given metric. Given node u, the set of neighbors is depicted by $\Gamma(u)$. The degree of node u is depicted as $k_u = |\Gamma(u)|$.

These metrics have the drawback that they do not account for the directed nature of follower networks. In other words, the metrics in Table 2.1 do not allow for differentiation whether a link will be established from, say, u to v or from v to u. As a next step we introduce patterns to account for directed links in social networks.

2.4.2 Triad Patterns

In social network theory a basic unit of analysis is a dyad. In undirected networks, a dyad is a pair of nodes who may share a social relation with one another. In directed networks, a dyad consists of a pair of nodes who may share a social relation through mutual links, an unreciprocated relation, or no relation. Unreciprocated means that one node is interested in the other node but not vice versa. A triad is a set of three parties, which consists of three dyads. A triad is "closed" if all nodes are linked with each other in some manner. A closed triad is also called triangle.

Figure 2.2 shows triad patterns of the actors u, z, and v. Edges are directed because our aim is to model patterns in directed social networks (e.g., follower networks). All patterns are open triads with z being the common neighbor of u and v. The questions with regards to link prediction can be stated as follows:

- What is the likelihood that u will establish a link towards v?
- In a partially observed network, is there a missing link pointing from u to v?

In Fig. 2.2 all possible connectivity configurations between u, z, and v are shown with the condition that u and v are not directly connected. In this work, open triads are labeled as T0X where X is the running index with X = [1, 9]. The pattern T01 shows the case where u and z as well as v and z are mutually connected. According to the theory of triadic closure, the chances are high that u will also connect to v (i.e., z's friends will likely become u's friends). In a follower network, the pair u and z and v and z would mutually follow each other. In T02, only u and z are

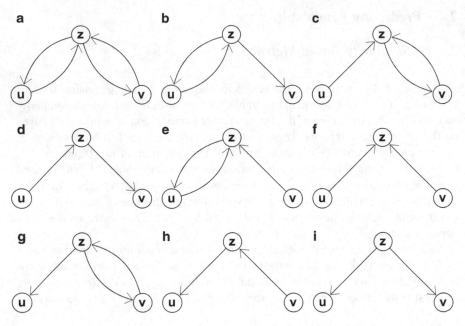

Fig. 2.2 Triad graph patterns. (**a**) T01. (**b**) T02. (**c**) T03. (**d**) T04. (**e**) T05. (**f**) T06. (**g**) T07. (**h**) T08. (**i**) T09

Fig. 2.3 Closed triads T1X. (**a**) T11. (**b**) T19

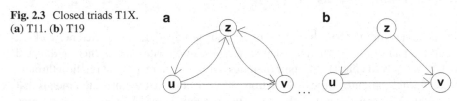

mutually connected to each other. The node v is followed by z but the relationship is not reciprocated. T03, T05, and T07 depict complementary cases where a mutual relation among one dyad exists. The other cases depicted by T04, T06, T08, and T09 show patterns without mutual relations among the dyads. The goal of link prediction is to determine which of the triads are or will be closed (i.e., becoming a triangle). A triad can be closed as follows if u establishes a link to v, v establishes a link to u, or if u and v establish a link mutually.

The following figures show closed triads based on T01 and T09 (the first and the last pattern of Fig. 2.2 are shown for brevity). Figure 2.3 shows the patterns where the triads are closed from u to v. The patterns are labelled similarly as in Fig. 2.2 but with a base offset of 10. Thus, the label of triads that are closed via u to v is T1X with $X = [1, 9]$.

Fig. 2.4 Closed triads T2X.
(a) T21. (b) T29

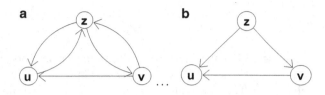

Fig. 2.5 Closed triads T3X.
(a) T31. (b) T39

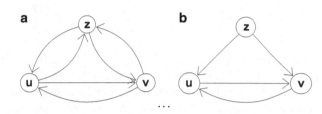

In the same manner, the label of triads that are closed via v to u is T2X with $X = [1,9]$ (see Fig. 2.4).

Finally, if the triads are closed by mutually connected u and v (see Fig. 2.5) the labels T3X with $X = [1,9]$ are applied. To summarize our discussions regarding triad patterns, triads that are relevant for link prediction may have 36 different configurations with regards to how nodes are connected to each other through directed links. Open triads have 2 connected dyads and have 2 to 4 links. Closed triads have 3 connected dyads and have 3 to 6 links.

The next step is to introduce a novel metric to calculate a score for link prediction based on the presented triad patterns.

2.4.3 Triadic Closeness

When considering a given pattern T0X (open triads as depicted by Fig. 2.2) the basic question is which of those patterns are likely closed triads (in the case of missing links) or which of those patterns will likely be closed in the future. Here we introduce *Triadic Closeness* (TC) to measure how close a pair of disconnected nodes are in terms of how the pair is connected through triads. TC is based on the following basic idea:

$$\text{Triadic Closeness} \propto \frac{\text{Number of closed triads}}{\text{Number of potentially closed triads}}$$

Triadic Closeness is thereby based on the ratio of the number of closed triads versus the number of potentially closed triads. Indeed, the chance that a given T0X triad will be closed depends on the actual social network and is most likely not the same for all follower networks. We define the TC score of the pair u and v in a directed graph G as follows:

$$TC_{uv} = \sum_{z \in \Gamma(u) \cap \Gamma(v)} w^P(u,v,z) \times w(z) \tag{2.1}$$

The score is calculated over all common neighbors. For a given neighbor z, the product is calculated by the triad weight $w^P(u,v,z)$ times the neighbor specific weight $w(z)$. The triad weight $w^P(u,v,z)$ is defined as follows:

$$w^P(u,v,z) = \frac{F(T(u,v,z)+10) + F(T(u,v,z)+30)}{F(T(u,v,z))} \tag{2.2}$$

The function $T(u,v,z)$ retrieves the triad pattern ID that matches the triad u, v, and z. The term $(T(u,v,z)+10)$ simply means that the ID of the closed triad counterpart of $T(u,v,z)$ is obtained (closed via u to v). Similarly, $(T(u,v,z)+30)$ gets the closed counterpart triad ID wherein $T(u,v,z)$ is closed through mutual links between u and v. The function $F(\cdot)$ retrieves the frequency of the given triad pattern. Prior to performing the calculation of $w^P(u,v,z)$, the frequencies of triads in a particular social network are computed by an algorithm and saved in a database. Afterwards, $F(\cdot)$ simply retrieves the triad frequency from a database (zero if the given triad was not detected in the graph).

The neighbor specific weight $w(z)$ can be tuned to account for the characteristics of specific social networks. For the basic case with $w(z) = 1$, TC_{uv} is only based on $w^P(u,v,z)$. We define the weight as $w(z) = \frac{1}{k_z}$ to give less connected neighbors more weight and thus TC_{uv} becomes:

$$TC_{uv} = \sum_{z \in \Gamma(u) \cap \Gamma(v)} w^P(u,v,z) \times \frac{1}{k_z} \tag{2.3}$$

Neighbors that are unique to only a few users are weighted more with $w(z) = \frac{1}{k_z}$ than popular neighbors. Popular neighbors are those with a high degree k_z and especially in networks such as Twitter these neighbors may be celebrities that may not have great significance for the triadic closure process. From a technical point of view, TC_{uv}'s behavior is comparable to Adamic-Adar Index (AA) [1] or the Resource Allocation Index (RA) [46] (see also Table 2.1). Indeed, the indices lack the notion of patterns and have been designed with undirected friendship networks in mind.

2.4.4 Triadic Closeness Example

To give a concrete example, consider the artificial network as depicted by Fig. 2.6 and suppose triadic closeness TC_{gh} shall be calculated between the nodes g and h.

The triad frequencies are listed in Table 2.2. Algorithm 2 shows the steps for counting the frequency of patterns in graph G.

Fig. 2.6 Example network

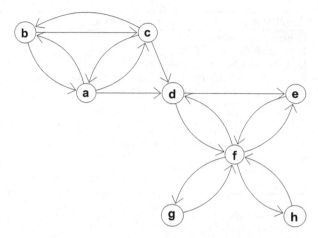

Table 2.2 Triads in example network

ID	Pattern	Frequency ↓
1	$u \leftrightarrow z \leftrightarrow v$	10
31	$u \leftrightarrow z \leftrightarrow v \leftrightarrow u$	6
2	$u \leftrightarrow z \rightarrow v$	2
3	$u \rightarrow z \leftrightarrow v$	2
4	$u \rightarrow z \rightarrow v$	2
5	$u \leftrightarrow z \leftarrow v$	2
7	$u \leftarrow z \leftrightarrow v$	2
8	$u \leftarrow z \leftarrow v$	2
12	$u \leftrightarrow z \rightarrow v \leftarrow u$	2
27	$u \leftarrow z \leftrightarrow v \rightarrow u$	2
36	$u \rightarrow z \leftarrow v \leftrightarrow u$	2
11	$u \leftrightarrow z \leftrightarrow v \leftarrow u$	1
21	$u \leftrightarrow z \leftrightarrow v \rightarrow u$	1
32	$u \leftrightarrow z \rightarrow v \leftrightarrow u$	1
33	$u \rightarrow z \leftrightarrow v \leftrightarrow u$	1
35	$u \leftrightarrow z \leftarrow v \leftrightarrow u$	1
37	$u \leftarrow z \leftrightarrow v \leftrightarrow u$	1

As shown in Fig. 2.6, the node f connects g and h via triad T01. Related to T01 for calculation are T11 and T31. The weight $w^P(u,v,z)$ for the network in Fig. 2.6 is given as $w^P(u,v,z) = \frac{6+1}{10} = 0.7$. Thus, TC_{gh} is given as $TC_{gh} = 0.7 \times \frac{1}{4} = 0.175$. Using the triadic closeness concept, with a probability of 0.17 T01 will be closed from g to h.

Algorithm 2 Pattern counting algorithm

1: **input:** directed graph G
2: **for each** Vertex $u \in G$ **do**
3:    **for each** Vertex $z \in getNeighbors(G,u)$ **do**
4:       **for each** Vertex $v \in getNeighbors(G,z)$ **do**
5:          **if** $equals(v,u)$ **then**
6:             continue
7:          **end if**
8:          $id \leftarrow getPatternId(G,u,z,v)$
9:          $count(id)$ // increment count by 1
10:       **end for**
11:    **end for**
12: **end for**

Fig. 2.7 Indegree
distribution of GitHub

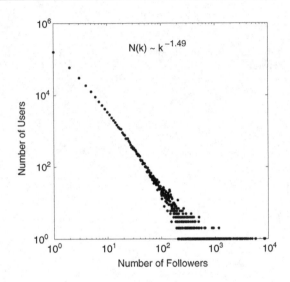

2.5   Data Collection

The following datasets have been used for testing the link prediction techniques.

- **GitHub.** The first network is based on GitHub's follower network [12]. The graph was imported in our prediction framework through the GitHub API [13] in December 2012. The basic network characteristic in terms of follower (indegree) distribution is depicted by Fig. 2.7.

  The plot shows a power-law distribution with the basic property $N(k) \sim k^{-1.49}$ where $N(k)$ is the number of nodes with indegree k. The follower graph counts 1,105,150 users and 1,898,034 following relations (edges). Nearly 70 % of users (767,975) have no followers (zero indegree) and again about 70 % of users (769,283) do not follow any other user (zero outdegree).

Fig. 2.8 Indegree
distribution of GooglePlus

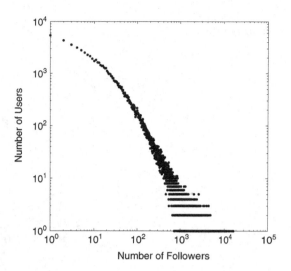

- **GooglePlus.** The second network represents a subset of nodes and edges from
  GooglePlus [14]. We have obtained the network (plain text files) from the
  Stanford Large Network Dataset Collection [41]. A description of the network
  is also given in [27]. The degree distribution is depicted by Fig. 2.8. The
  network consists of 107,614 nodes and 13,673,453 edges. At a technical level,
  the network is managed within the prediction framework's Flat File Store.
  The average degree $\langle k \rangle = 127$ in this network is much higher than in the
  GitHub-based follower network, which has only $\langle k \rangle = 1.7$.

  A possible explanation for the high differences in the average degree $\langle k \rangle$
  is the primary purpose of the platforms. GitHub is a platform for hosting and
  sharing source code repositories and the "follow" feature is by many people used
  to follow top-developers. In GooglePlus, many people follow other people they
  personally know and use the platform to maintain social relations.
- **Twitter.** The third network is based on a subset of nodes and edges of Twitter [43]. The network was also obtained from the Stanford Large Network Dataset
  Collection [41] and is also managed within the Flat File Store. The network
  counts 81,306 nodes and 1,768,149 edges, thereby making it the smallest network
  in our experiments. The degree distribution is depicted by Fig. 2.9. The average
  degree is given as $\langle k \rangle = 21.7$. In addition, we show the degree of a much
  larger Twitter-based network in Fig. 2.9 in the top-right corner to show how the
  presented network subset relates to the large network. The large network was
  obtained in July 2009 by [21] and counts roughly 5×10^7 nodes and 1.5×10^9
  edges.

  Both networks follow a similar distributional shape. In our experiments,
  only the smaller network has been used. In combination with the other larger
  networks (GitHub and GooglePlus), the smaller Twitter-based network provides
  a sufficient basis to compare link prediction metrics.

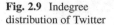

Fig. 2.9 Indegree
distribution of Twitter

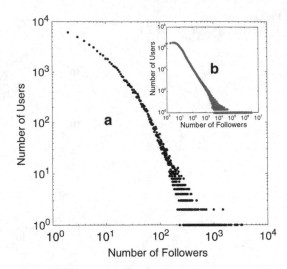

2.6 Evaluation

We obtained three directed social follower networks to compare different metrics
and to validate the suitability of TC. The networks (the whole follower network or
subsets thereof) include GitHub [12], GooglePlus [14], and Twitter [43].

2.6.1 Configuration

Here we discuss the link prediction configuration settings that were used to
perform experiments. Experiments have been performed using the three previously
introduced datasets: GitHub (**GH**), GooglePlus (**GP**), and Twitter (**TW**). Table 2.3
lists the prediction user set size $|U^P|$, the validation set size $|E^V|$ and the root set
size $|U^R|$.

For GitHub, the prediction user set U^P consists of 10% of the users which are
connected through 148,796 edges. From those edges we sampled 29,759 random
edges (20%) and added them to E^V. The root set U^R is populated with 100 nodes.
The size of the prediction target set U^T can be easily calculated as $U^T = U^P - U^R$. In
GooglePlus, we use the entire user base for U^P, which also results in approximately
the same size of U^P as for GitHub's prediction user set. E^V consists of 2,709,731
edges (20%) and the root set U^R consists also of 100 nodes. The Twitter-based
dataset has the smallest number of nodes and, in the same manner as for GooglePlus,
we also select all nodes for U^P. E^V has 333,577 edges (again 20%) and the same
root set size as for GitHub and GooglePlus is applied.

Using the configuration settings in Table 2.3, we obtain the graph $G^P(U, E^P)$
upon which triad pattern mining and prediction is performed. The relative frequency
of each triad pattern in G^P is listed in Table 2.4. The total number of triad
patters in each network is 423,034,532 (GitHub), 14,402,457,330 (GooglePlus), and

Table 2.3 Configuration
settings for link prediction

Configuration	GH	GP	TW
$|U^P|$	110,515	107,614	81,306
$|E^V|$	29,759	2,709,731	333,577
$|U^R|$	100	100	100

Table 2.4 Ratio of triads in
different social networks

ID	Pattern	GH (%)	GP (%)	TW (%)
9	$u \leftarrow z \rightarrow v$	53.23	9.97	7.04
6	$u \rightarrow z \leftarrow v$	39.09	18.79	39.96
8	$u \leftarrow z \leftarrow v$	1.50	12.71	5.95
4	$u \rightarrow z \rightarrow v$	1.37	12.71	5.95
7	$u \leftarrow z \leftrightarrow v$	1.22	5.69	4.36
2	$u \leftrightarrow z \rightarrow v$	1.14	5.69	4.36
5	$u \leftrightarrow z \leftarrow v$	0.63	3.93	6.21
3	$u \rightarrow z \leftrightarrow v$	0.63	3.93	6.21
1	$u \leftrightarrow z \leftrightarrow v$	0.23	1.44	4.07
28	$u \leftarrow z \leftarrow v \rightarrow u$	0.10	2.74	0.86
29	$u \leftarrow z \rightarrow v \rightarrow u$	0.10	2.74	0.86
19	$u \leftarrow z \rightarrow v \leftarrow u$	0.10	2.74	0.86
26	$u \rightarrow z \leftarrow v \rightarrow u$	0.10	2.74	0.86
16	$u \rightarrow z \leftarrow v \leftarrow u$	0.09	2.74	0.86
14	$u \rightarrow z \rightarrow v \leftarrow u$	0.09	2.74	0.86
25	$u \leftrightarrow z \leftarrow v \rightarrow u$	0.05	0.62	0.70
39	$u \leftarrow z \rightarrow v \leftrightarrow u$	0.05	0.62	0.70
13	$u \rightarrow z \leftrightarrow v \leftarrow u$	0.04	0.62	0.70
27	$u \leftarrow z \leftrightarrow v \rightarrow u$	0.04	1.41	0.87
12	$u \leftrightarrow z \rightarrow v \leftarrow u$	0.04	1.41	0.87
36	$u \rightarrow z \leftarrow v \leftrightarrow u$	0.04	1.41	0.87
31	$u \leftrightarrow z \leftrightarrow v \leftrightarrow u$	0.03	0.29	0.97
11	$u \leftrightarrow z \leftrightarrow v \leftarrow u$	0.01	0.21	0.53
33	$u \rightarrow z \leftrightarrow v \leftrightarrow u$	0.01	0.21	0.53
21	$u \leftrightarrow z \leftrightarrow v \rightarrow u$	0.01	0.21	0.53
37	$u \leftarrow z \leftrightarrow v \leftrightarrow u$	0.01	0.21	0.53
32	$u \leftrightarrow z \rightarrow v \leftrightarrow u$	0.01	0.21	0.53
35	$u \leftrightarrow z \leftarrow v \leftrightarrow u$	0.01	0.21	0.53
17	$u \leftarrow z \leftrightarrow v \leftarrow u$	0.00	0.15	0.27
23	$u \rightarrow z \leftrightarrow v \rightarrow u$	0.00	0.15	0.27
22	$u \leftrightarrow z \rightarrow v \rightarrow u$	0.00	0.15	0.27
38	$u \leftarrow z \leftarrow v \leftrightarrow u$	0.00	0.15	0.27
15	$u \leftrightarrow z \leftarrow v \leftarrow u$	0.00	0.15	0.27
34	$u \rightarrow z \rightarrow v \leftrightarrow u$	0.00	0.15	0.27
18	$u \leftarrow z \leftarrow v \leftarrow u$	0.00	0.10	0.13
24	$u \rightarrow z \rightarrow v \rightarrow u$	0.00	0.10	0.13

296,075,286 (Twitter). The patterns at the top in Table 2.4 are open triads T01–T09 followed by closed triads T11–T36. The patterns are sorted in descending order by the column **GH**.

The most common triad pattern in GitHub with 53.23 % is where both u and v are followed by z but neither u nor v follow z. This pattern has a relative frequency of 7.04 % in Twitter, thereby being the second most common pattern in Twitter, and a relative frequency of 9.97 % in GooglePlus, thereby being the fourth most common pattern in GooglePlus. The second most common pattern in GitHub, and the most common pattern in GooglePlus and Twitter, is the pattern where both u and v follow z but z follows neither of them. Triadic Closeness is defined as the likelihood that a given open triad (T01–T09) will be closed in a given social network. Thus, the triad patterns T01–T09 are seen in relation with the closed triads to determine the closeness between two nodes.

In the following the prediction results are presented by plotting ROC curves and calculating AUC for each metric. TC is based on the frequencies of respective triads in Table 2.4.

2.6.2   Prediction Results

In all our experiments, we use the metrics as defined in Table 2.1 and the introduced triadic closeness TC. The configuration settings for the experiments are given in Table 2.3. As a general note, an AUC value above 0.5 indicates that a prediction algorithm performs better than pure chance. Higher AUC indicates better prediction accuracy. With regards to ROC curves, we group metrics into a single *class* if their AUC values are identical.

GitHub As a first step we present the prediction results of GitHub-based experiments. The ROC curves are depicted by Fig. 2.10. The metrics and class correspondence is established in Table 2.5. We provide the AUC values along with the curves in Fig. 2.10 and also in Table 2.5 for easier readability.

In GitHub, TC corresponds to C1 and an AUC value of 0.98. Thus, TC outperforms the other methods and delivers the highest accuracy. RA delivers also very good results with an AUC of 0.97. Metrics below 0.5 such as SA, LHN, JA, SO, and HD are not suitable prediction methods for the GitHub-based social networks. The HitRatio at different thresholds is shown in Table 2.6. The HitRatio until HitRatio@50 is identical for CN, AA and TC. For HitRatio@100 and HitRatio@1000 TC outperforms other methods. RA performs best at HitRatio@5000.

GooglePlus Next, we present the prediction results of GooglePlus-based experiments. The ROC curves are presented in Fig. 2.11 and the class correspondence is given in Table 2.7. TC has an AUC value of 0.96 followed by RA and AA with 0.94. Also, the simple CN metric delivers quite good results with an AUC of 0.93. LHN perform worst but still has an AUC above 0.5 thereby delivering acceptable results but with low accuracy.

Fig. 2.10 ROC curves for
GitHub-based results

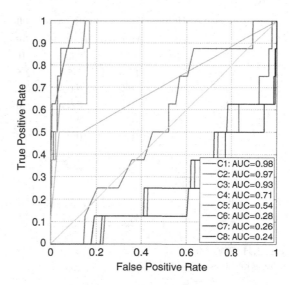

Table 2.5 AUC Classes for
GitHub-based results

Class	Definition	AUC
C1	TC	0.98
C2	RA	0.97
C3	AA	0.93
C4	CN	0.71
C5	HP	0.54
C6	SA	0.28
C7	LHN	0.26
C8	JA, SO, HD	0.24

The HitRatio for GooglePlus-based results is shown in Table 2.8. TC performs best at all thresholds and delivers the best results compared with the other methods. RA performs slightly better than AA in terms of HitRatio. LHN performed worst in terms of AUC but slightly better than HP with regards to HitRatio. Overall, only TC, RA and AA are suitable methods to perform link prediction in GooglePlus. All other methods have no correct results at HitRatio@30.

Twitter Finally, we present the prediction results of Twitter-based experiments. The ROC curves are presented in Fig. 2.12 and the class correspondence is given in Table 2.9.

Again, TC performs best with an AUC value of 0.97. Second ranked are again RA and AA with an AUC of 0.96. Note in this context that all three metrics, TC, RA, and AA, give higher weights to those neighbors who have a lower degree (i.e., "hub-depressed" behavior). However, by considering triad patterns TC outperforms all other metrics and delivers the most accurate results. Again, LHN ranks last with an AUC of 0.71. None of the metrics have an AUC lower than 0.5. Here the lowest

Table 2.6 HitRatio (%) for GitHub-based results

Metric	HitRatio @10	HitRatio @30	HitRatio @50	HitRatio @100	HitRatio @1000	HitRatio @5000
CN	12.5	12.5	12.5	12.5	12.5	50.0
SA	0.0	0.0	0.0	0.0	0.0	0.0
JA	0.0	0.0	0.0	0.0	0.0	0.0
SO	0.0	0.0	0.0	0.0	0.0	0.0
HP	0.0	0.0	0.0	0.0	0.0	0.0
HD	0.0	0.0	0.0	0.0	0.0	0.0
LHN	0.0	0.0	0.0	0.0	0.0	0.0
AA	12.5	12.5	12.5	12.5	37.5	62.5
RA	0.0	0.0	12.5	12.5	37.5	87.5
TC	12.5	12.5	12.5	37.5	62.5	75.0

Fig. 2.11 ROC curves for GooglePlus-based results

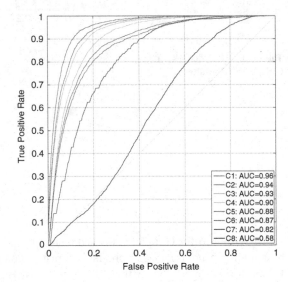

AUC is 0.71 making all metrics suitable methods for link prediction. As mentioned before, this was not the case for GitHub where many metrics perform below an AUC of 0.5.

The HitRatio for Twitter-based results is shown in Table 2.10. TC performs best until HitRatio@100. For HitRatio@1000 and HitRatio@5000 RA performs slightly better. However, TC still performs best with respect to AUC and true positive rate. Also, the simple common neighbor (CN) methods provides acceptable results in terms of HitRatio. LHN performs worst with regards to HitRatio (only 5.2 % at HitRatio@5000) and also with regards to AUC.

Table 2.7 AUC Classes for
GooglePlus-based results

Class	Definition	AUC
C1	TC	0.96
C2	RA, AA	0.94
C3	CN	0.93
C4	SA	0.90
C5	JA, SO	0.88
C6	HD	0.87
C7	HP	0.82
C8	LHN	0.58

Table 2.8 HitRatio (%) for GooglePlus-based results

Metric	HitRatio @10	HitRatio @30	HitRatio @50	HitRatio@ @100	HitRatio@ @1000	HitRatio @5000
CN	0.0	0.0	0.1	0.3	2.3	8.1
SA	0.0	0.0	0.0	0.0	0.1	1.4
JA	0.0	0.0	0.0	0.0	0.5	1.7
SO	0.0	0.0	0.0	0.0	0.1	1.5
HP	0.0	0.0	0.0	0.0	0.0	0.0
HD	0.0	0.0	0.0	0.0	0.1	1.7
LHN	0.0	0.0	0.0	0.0	0.0	0.2
AA	0.0	0.1	0.1	0.4	2.4	8.6
RA	0.1	0.2	0.3	0.6	3.4	12.8
TC	0.3	0.7	0.8	1.1	6.0	21.6

Fig. 2.12 ROC curves for
Twitter-based results

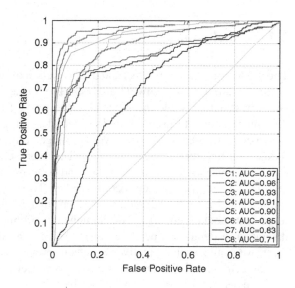

Table 2.9 AUC Classes
Twitter-based results

Class	Definition	AUC
C1	TC	0.97
C2	RA, AA	0.96
C3	CN	0.93
C4	HP	0.91
C5	SA	0.90
C6	JA, SO	0.85
C7	HD	0.83
C8	LHN	0.71

Table 2.10 HitRatio (%) for Twitter-based results

Metric	HitRatio @10	HitRatio @30	HitRatio @50	HitRatio @100	HitRatio @1000	HitRatio @5000
CN	2.6	6.1	9.2	13.1	44.1	72.1
SA	0.0	0.0	1.3	5.2	28.4	56.8
JA	1.7	3.1	3.1	7.9	34.1	56.8
SO	0.0	0.0	1.3	4.4	29.3	54.6
HP	0.0	0.0	0.0	0.0	0.4	40.2
HD	0.9	2.6	2.6	2.6	25.3	49.8
LHN	0.0	0.0	0.0	0.0	0.0	5.2
AA	2.6	6.1	9.6	14.4	48.9	78.6
RA	3.1	6.6	8.3	12.2	56.3	86.5
TC	3.9	8.7	11.8	14.8	55.5	86.4

2.7 Conclusions

The prediction of missing links and the prediction of future links is an important task in the domain of social network analysis. The former helps to infer the "real" social network structure while the latter is used to give friendship as well as following recommendations to users. A wide range of local, global, and semi-local metrics have been proposed by previous work. A large body of existing literature, however, focuses on undirected networks only. This work closes this gap by focusing on directed networks. Here we propose triad patterns to predict links between nodes in directed graphs. Our approach is called Triadic Closeness. We designed and implemented a link prediction framework that is able to perform predictions in large-scale social networks. The framework's architecture has been presented and discussed in detail. We performed experiments in three different social networks. First, we analyzed the effectiveness of our proposed approach in GitHub; a social coding community. Second, we obtained a subset of the GooglePlus network and third we performed experiments in a subset of the Twitter follower network.

- The follower structure of GitHub and GooglePlus or Twitter is very different. Thus, the average degree of GitHub is significantly lower than the average degree in GooglePlus and Twitter.
- As expected, each follower network exhibits distinct triad frequencies. The presented approach helps to give higher weights to triads that are more frequently closed (resulting in closed triangles).
- The pattern-based prediction approach delivers the best results among the compared local methods. TC consistently outperformed other approaches. Thus, a pattern-based approach is better suited in directed social networks.

An important aspect will also be the extension of our approach towards global and semi-local methods. As an example, the personalized PageRank method could provide the basis for pattern-aware link prediction. We are currently working on the design of this method. In addition, we will be comparing machine learning techniques for link prediction such as random forest and support vector machine (SVM) models.

References

1. L. A. Adamic and E. Adar. Friends and neighbors on the web. *Social Networks*, 25:211–230, 2001.
2. L. M. Aiello, A. Barrat, R. Schifanella, C. Cattuto, B. Markines, and F. Menczer. Friendship prediction and homophily in social media. *ACM Trans. Web*, 6(2):9:1–9:33, June 2012.
3. E. M. Airoldi, D. M. Blei, S. E. Fienberg, and E. P. Xing. Mixed membership stochastic blockmodels. *J. Mach. Learn. Res.*, 9:1981–2014, June 2008.
4. U. Alon. Network motifs: theory and experimental approaches. *Nature Reviews Genetics*, 8(6):450–461, June 2007.
5. L. Backstrom and J. Leskovec. Supervised random walks: predicting and recommending links in social networks. In *Proceedings of the fourth ACM international conference on Web search and data mining*, WSDM '11, pages 635–644, New York, NY, USA, 2011. ACM.
6. V. Batagelj and A. Mrvar. A.: A subquadratic triad census algorithm for large sparse networks with small maximum degree. *Social Networks*, pages 237–243, 2001.
7. A. P. Bradley. The use of the area under the roc curve in the evaluation of machine learning algorithms. *Pattern Recognition*, 30:1145–1159, 1997.
8. M. J. Brzozowski and D. M. Romero. Who should i follow? recommending people in directed social networks. In L. A. Adamic, R. A. Baeza-Yates, and S. Counts, editors, *ICWSM*. The AAAI Press, 2011.
9. A. Clauset, C. Moore, and M. E. J. Newman. Hierarchical structure and the prediction of missing links in networks. *Nature*, 453(7191):98–101, May 2008.
10. I. Esslimani, A. Brun, and A. Boyer. Densifying a behavioral recommender system by social networks link prediction methods. *Social Netw. Analys. Mining*, 1(3):159–172, 2011.
11. Facebook. Online: www.facebook.com (last access June 2015).
12. GitHub. Online: www.github.com (last access June 2015).
13. GitHub. Online: http://developer.github.com/ (last access June 2015).
14. GooglePlus. Online: www.plus.google.com/ (last access June 2015).
15. M. Granovetter. The strength of weak ties. *The American Journal of Sociology*, 78(6): 1360–1380, 1973.

16. P. W. Holland, K. B. Laskey, and S. Leinhardt. Stochastic blockmodels: First steps. *Social Networks*, 5(2):109–137, June 1983.
17. P. W. Holland and S. Leinhardt. A method for detecting structure in sociometric data. *American Journal of Sociology*, 76(3):492–513, 1970.
18. G. Jeh and J. Widom. Simrank: a measure of structural-context similarity. In *Proceedings of the eighth ACM SIGKDD international conference on Knowledge discovery and data mining*, KDD '02, pages 538–543, New York, NY, USA, 2002. ACM.
19. G. Jeh and J. Widom. Scaling personalized web search. In *Proceedings of the 12th international conference on World Wide Web*, WWW '03, pages 271–279, New York, NY, USA, 2003. ACM.
20. L. Katz. A new status index derived from sociometric analysis. *Psychometrika*, 18(1):39–43, Mar. 1953.
21. H. Kwak, C. Lee, H. Park, and S. Moon. What is twitter, a social network or a news media? In *Proceedings of the 19th international conference on World wide web*, WWW '10, pages 591–600, New York, NY, USA, 2010. ACM.
22. E. A. Leicht, P. Holme, and M. E. J. Newman. Vertex similarity in networks. *Phys. Rev. E*, 73:026120, Feb 2006.
23. J. Leskovec, D. Huttenlocher, and J. Kleinberg. Predicting positive and negative links in online social networks. In *Proceedings of the 19th international conference on World wide web*, WWW '10, pages 641–650, New York, NY, USA, 2010. ACM.
24. D. Liben-Nowell and J. Kleinberg. The link prediction problem for social networks. In *Proceedings of the twelfth international conference on Information and knowledge management*, CIKM '03, pages 556–559, New York, NY, USA, 2003. ACM.
25. W. Liu and L. Lu. Link prediction based on local random walk. *EPL (Europhysics Letters)*, 89(5):58007, 2010.
26. L. Lu and T. Zhou. Link prediction in complex networks: A survey. *Physica A: Statistical Mechanics and its Applications*, 390(6):1150–1170, 2011.
27. J. J. McAuley and J. Leskovec. Learning to discover social circles in ego networks. In P. L. Bartlett, F. C. N. Pereira, C. J. C. Burges, L. Bottou, and K. Q. Weinberger, editors, *NIPS*, pages 548–556, 2012.
28. B. Meng, H. Ke, and T. Yi. Link prediction based on a semi-local similarity index. *Chinese Physics B*, 20(12):128902, 2011.
29. R. Milo, S. Shen-Orr, S. Itzkovitz, N. Kashtan, D. Chklovskii, and U. Alon. Network motifs: Simple building blocks of complex networks. *Science*, 298(5594):824–827, Oct. 2002.
30. L. Page, S. Brin, R. Motwani, and T. Winograd. The pagerank citation ranking: Bringing order to the web, 1999.
31. E. Ravasz, A. L. Somera, D. A. Mongru, Z. N. Oltvai, and A. L. Barabasi. Hierarchical organization of modularity in metabolic networks. *Science*, 297(5586):1551–1555, Aug. 2002.
32. A. Rettinger, H. Wermser, Y. Huang, and V. Tresp. Context-aware tensor decomposition for relation prediction in social networks. *Social Netw. Analys. Mining*, 2(4):373–385, 2012.
33. D. M. Romero and J. M. Kleinberg. The directed closure process in hybrid social-information networks, with an analysis of link formation on twitter. In W. W. Cohen and S. Gosling, editors, *ICWSM*. The AAAI Press, 2010.
34. G. Salton and M. J. McGill. *Introduction to Modern Information Retrieval*. McGraw-Hill, Inc., New York, NY, USA, 1986.
35. G. Sautter and K. BâĂŽĂăĂűâĂŽĂăĂĞhm. High-throughput crowdsourcing mechanisms for complex tasks. *Social Network Analysis and Mining*, pages 1–16, 2013.
36. D. Schall. Expertise ranking using activity and contextual link measures. *Data Knowl. Eng.*, 71(1):92–113, 2012.
37. D. Schall. *Service Oriented Crowdsourcing: Architecture, Protocols and Algorithms*. Springer Briefs in Computer Science. Springer New York, New York, NY, USA, 2012.
38. D. Schall and F. Skopik. Social network mining of requester communities in crowdsourcing markets. *Social Netw. Analys. Mining*, 2(4):329–344, 2012.
39. T. A. Snijders. *Transitivity and Triads*. University of Oxford, 2012. Online: http://www.stats. ox.ac.uk/~snijders/Trans_Triads_ha.pdf (last access 22-Feb-2013).

40. T. Sørensen. A method of establishing groups of equal amplitude in plant sociology based on similarity of species and its application to analyses of the vegetation on danish commons. *Biologiske Skrifter / Kongelige Danske Videnskabernes Selskab*, 5(4):1–34, 1957.
41. Stanford. Online: http://snap.stanford.edu/data/index.html (last access January 2014).
42. P. Symeonidis and N. Mantas. Spectral clustering for link prediction in social networks with positive and negative links. *Social Network Analysis and Mining*, pages 1–15, 2013.
43. Twitter. Online: www.twitter.com (last access June 2015).
44. S. Wasserman, K. Faust, and D. Iacobucci. *Social Network Analysis : Methods and Applications (Structural Analysis in the Social Sciences)*. Cambridge University Press, Nov. 1994.
45. H. C. White, S. A. Boorman, and R. L. Breiger. Social structure from multiple networks. i. blockmodels of roles and positions. *The American Journal of Sociology*, 81(4):730–780, 1976.
46. T. Zhou, L. Lu, and Y.-C. Zhang. Predicting missing links via local information. *The European Physical Journal B - Condensed Matter and Complex Systems*, 71(4):623–630, Oct. 2009.

Chapter 3
Follow Recommendation in Communities

Abstract Follower networks provide means for informal information propagation. In this chapter we introduce an approach for recommending relevant users to follow. Our approach is based on the automatic analysis of user behavior and network structure. Link-analysis techniques such as PageRank and HITS provide the basis for a novel recommendation model.

3.1 Social Collaboration Platforms

Social networks have become a central part for many people in their everyday activities. The type of network used for different activities often varies depending on the desired purpose. Professional networks such as LinkedIn are used to stay in touch with colleagues and coworkers. Personal social networks including the popular Facebook platform enable people to engage with their friends and to follow their news updates. News media and social network services such as Twitter allow people to follow short news updates (tweets) of celebrities and friends. Recently, another type of social network has become highly popular attracting millions of users: *online social collaboration networks*. These networks enable people to collectively work on projects. An example of such a social collaboration platform is Github [1]. GitHub was launched in 2008 and enables people to work on public (open source) or private projects. Indeed, open source development has a long history (e.g., see [2]) and dates back to the 1950s and 1960s when IBM released software sources of its operating systems and other programs [3].

A novel aspect of recent online social collaboration platforms on the World Wide Web such as GitHub is that they provide improved support for social networking features such as followers/followings or news feeds based on the users' social network. These online collaboration platforms enable users to discover interesting projects and repositories more quickly and let people collaborate almost instantaneously. An intriguing hypothesis was postulated by [4] arguing that GitHub will be the next big social network driven by what people do instead of who people know. In professional networks such as LinkedIn people are mainly connected because they know each other from, for example, past work experience.

In networks such as LinkedIn or Facebook friendship is represented as reciprocated links in an undirected graph. Services such as Twitter and recently GitHub are based on a directed network approach. A directed network approach allows users to follow other users based on their interest without requiring them to reciprocate the relationship. In traditional social networks, some users may be followed by many people without following many peers themselves ("stars" or "celebrities"). Is this also the case for online social collaboration networks such as GitHub? People in GitHub are mostly followed because they work on interesting projects. Thus, this difference between conventional social networks and online social collaboration networks requires a novel follow recommendation approach. One important aspect in knowledge-intensive disciplines such as software engineering is to promote the effective dissemination of knowledge [5]. The authors in [6] found that formal routines should be supplemented by collaborative and social processes to promote awareness and learning. In our opinion, follower networks provide excellent means to address the need for effective dissemination of knowledge through informal relations and information interest. Following the right person is essential to get information updates from the community leaders and "gurus". Follow recommendations aim to solve the problem of selecting the right person to follow.

Follower networks, information flows through re-tweets, and follow recommendations have been analyzed in great detail for platform such as Twitter [7–9] or in enterprise social media networks such as WaterCooler [10]. To our best knowledge, there is no existing work that proposed context-sensitive following recommendations in online development networks.

In this chapter we present the following key contributions:

- **Who to follow recommendation.** Here we propose a method and algorithm for follow recommendations. Recommendations can be based on *behavior*, *network*, or *similarity*. Our approach is based on network analysis techniques. User relevance with respect to following recommendations is based on a novel authority metric. Authority in this work means being an expert or guru in a specific area (e.g., expert/guru in "javascript" programming). The approach is specifically targeted at online software development communities but may be applied to other types of collaboration networks as well.
- **Social network metrics and evaluation.** We analyze our approach by using social network (follower network) and activity data from GitHub. We introduce the used dataset and calculate various metrics such as reciprocity to support our hypothesis that people in GitHub are mostly followed because they work on interesting projects. The proposed authority-based follow recommendation approach is evaluated by using the GitHub dataset.

This chapter is structured as follows: Sect. 3.2 discusses relevant related work. Section 3.4 introduces our follow recommendation approach. The data collection is introduced in Sect. 3.5. Section 3.6 introduces social network metrics and our evaluation. The chapter is concluded in Sect. 3.7 with an outlook on future research.

3.2 Background in Online Communities

We structure related work into relevant topics including analysis of online development communities, and social network analysis.

Online Development Communities An online social network is a communication and collaboration medium that connects a large number of people. People within the social network stay together if their interaction dynamics leads to the emergent property that is called "community". Here we focus on online development communities consisting of people developing collaboratively open source projects. A topological analysis of the SourceForge community was presented in [11]. The focus of the work was on role detection of users and cluster analysis. In [12], metrics with regards to open versus private software development were analyzed with the focus on source code aspects. Measures to investigate the social-technical congruence in software development projects were established in [13]. The interplay between network metrics, software failures and software evolution was investigated in, for example [14–16]. Collaboration and influence on GitHub was analyzed in [17] with the central focus on visualization techniques. Interesting directions with regards to the analysis of GitHub were presented in [18]. The authors [18] showed evidence for social collaboration on GitHub and proposed algorithms addressing the team formation problem. Our authority-based recommendation approach can well be used to create expertise profiles that can be used to assist in the formation of expert teams [19, 20]. From a technical point of view, the basic structure of the GitHub API and the event schema was described in [21].

In this work, we analyze the GitHub online development community but focus on follower/following recommendations.

Social Network Analysis Social network analysis techniques offer a rich set of theories (e.g., social network theory, small world phenomenon, power-laws, self-organization, and graph theory) and tools to analyze the topological features and human behavior in online communities [7, 22–27].

In many systems, including large-scale enterprises, mostly searchable directories or databases that include descriptions of the employees' knowledge and experience are used to locate experts. The problem with this approach is that social networks and also big companies are in a constant state of flux and change [28, 29]. In large-scale online communities and dynamic organizations, it becomes infeasible to constantly review and update the profile information of often rapidly changing experience, skills, and rolls of experts. Specifically with respect to software engineering, the authors in [30] found that a person with the most modifications to the code may be an expert within a community, and that expertise depends on the area of the code that is being modified. Furthermore, the "degree of knowledge model" [31], showed how that expertise decays with subsequent changes by other authors.

We apply well-grounded theories and algorithms to the analysis of large scale software development communities. Well known ranking algorithms to calculate importance in linked environments include the HITS algorithm [32] and PageRank [33]. Personalization techniques such as topic-sensitive PageRank [34, 35] enable context-sensitive importance ranking. Link analysis algorithms have been successfully applied to estimate actor importance in social and collaborative networks [27, 36–38] as well as to online mass production systems such as emerging crowdsourcing environments [39]. The authors in   [37] proposed link analysis techniques such as PageRank for expertise mining in online communities. However, no personalization with regards to expertise areas has been performed by prior work. In our previous work [27] we proposed context-sensitive link analysis algorithms for expertise mining, but did not consider contributions of people to online communities (e.g., contributing source code to online repositories).

We propose a network-based metric to capture authority for follow recommendations. The proposed authority metric measures the relative community standing (i.e., reputation) of a community member. The metric is based on how much a person contributes to the community (e.g., repository commits) and on how many existing follower relations a person has.

3.3   Recommendation Types

Following/follower recommendations can be performed according to different strategies. The authors in [10] suggested three categories: behavioral, similarity, and network. We structure recommendation types for online software development communities in a similar manner but provide more strategies for network based recommendations. The following recommendation types can be distinguished:

- **Behavioral.** This recommendation type is based on already observed interaction between two people. For example, a person may have commented on code checked in by some other person or a person may have replied to a question (e.g., usage of library, report of bug, etc.) posted by someone in a discussion forum. Thus, based on the interaction between two people a follow recommendation can be made.
- **Similarity.** As often observed in real life, people tend to have friends with similar characteristics and interests [40]. This phenomenon is called homophily. Homophily is the principle that a contact between similar people occurs at a higher rate than among dissimilar people [41]. Shared characteristics include follower degree and shared interests include watched repositories or interest in programming language.

- **Network.** The network based recommendation type can be based on various techniques.

  Collaborative filtering [42] is a technique for recommending content to users based on other users with similar interest. Collaborative filtering techniques are commonly seen on e-commerce sites. Applied to our problem domain, if A watches the same software repositories as B, than A might be interested in following B because both share similar interests.

  Triadic closure [41] is another concept for network based recommendation. Suppose three community members A, B, and C and social ties between A-C and B-C. As suggested by Granovetter, in most of these social structures, a triadic closure occurs such that A and B are likely to become friends (or connected to each other) the more they associate with C [41].

  Network centrality of various types of vertices in a graph can be computed to determine the relative importance of vertices (e.g., methods such as PageRank [33] and HITS [32]). In social networks, network centrality techniques can be used to estimate the importance of users [27]. The application is, for example, expert recommendation.

In the scope of this work, we focus on the *network based recommendation type*. The first type (behavioral) and the second type (similarity) are not within the scope of this work. With regards to network based recommendation, we focus on a centrality based approach.

3.4 Follow Recommendation

3.4.1 Authority-Based Recommendation Model

The central concept in this work is *user authority*. Here we follow a social network analysis approach that is based on well established techniques including HITS [32] and PageRank [33]. In this work, somebody is considered to be an authority if s/he has knowledge in a given area and is recognized by the community. These two factors are combined in a novel authority metric. The presented techniques and algorithms build upon our previous work [27, 39]. Our previous work introduced a social network mining framework and context-sensitive expertise ranking algorithms. Here we introduce new metrics suitable for online development communities such as GitHub. Here, the basic follow recommendation model is established upon a user-repository graph model (in contrast to our previous work in [27]).

To illustrate the basic idea of our follow approach, without going into details, we show the most fundamental steps in Algorithm 3.

The input of the algorithm is a person *P* for whom follow recommendations shall be computed. Thus, a personalization procedure to compute recommendations is needed. The output of the algorithm is a top-*k* list of people to follow. The threshold *k* can be dynamically adjusted. As a first step, a query *Q* is created from the person

---

Algorithm 3 High level follow recommendation algorithm

1: **input:** Person P for whom recommendations should be computed.
2: **output:** Top-k list of recommended people.
3: $Q \leftarrow createQueryFromProfile(P)$
4: $D \leftarrow getPrincipleInterestDomains(P)$
5: **for each** User $u \in U$ **do**
6:    **if** $matches(u, D)$ **then**
7:       $A(u) \leftarrow computeAuthorityScore(u, Q)$
8:    **end if**
9: **end for**
10: **return** $sortAndFilter(A)$

P's profile (function $createQueryFromProfile()$ in line 3). As an example, if person P is interested in "javascript" and "closure", then s/he may want to follow people who work with such languages. The query would therefore consist of the keywords $Q = \{$"*javascript*", "*closure*"$\}$. The detailed mechanisms for extracting the query Q from P's user profile is not detailed in this work.

The next step is to extract principle interest domains for person P. Principle interest domains are, for example, "web engineering", "automation", "embedded systems development", "physics simulations". A person with background web engineering and interest in javascript-based web frameworks for e-commerce may or may not be interested in following a person with background "automation" who is working in javascript-based industry monitoring frameworks. The application domains can be very versatile and it may also strongly depend on the person's interest to follow somebody from another domain. Here, we highlight the fact that such filtering (see $matches(u, D)$ in line 6 which evaluates to true or false) may be needed to improve recommendations. However, it would be beyond the scope of this work to elaborate on filtering by principle interest domains in detail.

The core focus of this work is depicted by the function $computeAuthorityScore()$. Here the authority scores are computed that are used to perform a ranking of recommended users to follow. The function $sortAndFilter()$ truncates the list (if desired) and returns a sorted list of people (sorted by highest to lowest authority score).

Models to compute authority such as the popular hubs and authority approach [32] are rooted in the link analysis domain to rank Web pages. Here we apply link analysis mechanisms to evaluate user authority with respect to community contributions. Detecting authoritative users is not only important for follow recommendations but also for identifying key players in the community. These users have strong impact with regards to community cohesion and evolution. The following sections details the context-sensitive authority ranking model.

Fig. 3.1 User-repository
graph model

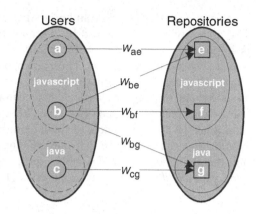

3.4.2   *Personalized Authority Ranking*

Here we define the follow recommendation model that is based on link analysis techniques. The basic idea is to compute user authority based on performed repository actions (e.g., coding activities, bug fixing, etc.). The concepts are depicted by Fig. 3.1.

Suppose the community consists of the set of users $U = \{a,b,c\}$ and the set of software repositories $R = \{e,f,g\}$. The user-repository graph can be modeled as a directed bipartite graph $G_B(V_U, V_R, E_B)$ where V_U represents the set of users and V_R the set of repositories. An edge $(u,r) \in E_B$ is established from u to r iff u performs an action in r. An edge between u and r is weighted by w_{ur} based on performed actions. Each repository is associated with a programming language. In Fig. 3.1, e and f are associated with *javascript* and g is associated with *java*. A user gains experience in a programming language if the user performs actions in language-related repositories. The programming languages denote the *context* for our personalized authority ranking approach.

The set of users $\{a,b\}$ has gained experience in *javascript* because they have performed actions in the repositories e and f whose language is *javascript*. The user b is experienced in *javascript* and *java* because b has performed actions in $\{e,f,g\}$. The users a and c are only experience in *javascript* and *java*, respectively.

A quite natural and intuitive approach to rank users and repositories in G_B is to model importance using the notion of *hubs* and *authorities*[1] as introduced in [32]:

$$A(u) = \sum_{(u,r)\in E_B} H(r) \qquad H(r) = \sum_{(v,r)\in E_B} A(v) \qquad (3.1)$$

---

[1]The algorithm introduced by Kleinberg [32] to compute the scores is called Hyperlink Induced Topic Search (HITS).

A user $u \in V_U$ has high authority if u contributes to important repositories. The authority of u is depicted by the authority score $A(u)$. A repository $r \in V_R$ is important if authoritative users contribute to it. The repository importance is depicted by the hub score $H(r)$. Thus, an important "hub" attracts many authoritative users. This recursive definition of user and repository importance provides the basis for our ranking approach. Other centrality metrics such as PageRank [33] cannot discriminate between two types of scores. However, when compared with PageRank, HITS is less stable to small perturbations [43]. Thus, a combined model would bring the advantage that one can compute two scores (property of HITS) and that the algorithm is rank stable (property of PageRank). Here we follow the *randomized HITS* approach as proposed in [43]:

$$A(u) = (1 - \lambda_a)p(u) + \lambda_a \sum_{(u,r) \in E_B} H(r) \tag{3.2}$$

$$H(r) = (1 - \lambda_h)p(r) + \lambda_h \sum_{(v,r) \in E_B} A(v) \tag{3.3}$$

Equations (3.2) and (3.3) show a natural way of designing a random-walk based algorithm following the HITS model. However, the randomized HITS approach is, like PageRank, stable to small perturbations [43]. The symbols $p(u)$ and $p(r)$ depict personalization vectors that may be assigned uniformly for each node such that $p(u) = \frac{1}{|V_U|}$ and $p(r) = \frac{1}{|V_R|}$.

Non-uniform personalization vectors result in personalized rankings (cf. personalized PageRank [33]). The parameters λ_a and λ_h with $0 \leq \lambda \leq 1$ allow for balancing between authority/hub weights and personalization weights. A typical value for λ is 0.85 [33]. Assigning lower values to λ means that higher importance is given to the personalization weights; thereby reducing the "*network effect*" of the ranking algorithm.

The idea of our personalized authority ranking approach is to compute ranking scores with respect to certain *interest areas*. The demanded areas of interest are passed via the keyword based query $Q = \{q_1, q_2, \ldots, q_n\}$ to the ranking algorithm. Each query keyword q_n corresponds to a desired area of interest. An interest area is identified via the name of a programming language (for example, $Q = \{$"*javascript*", "*java*"$\}$). A query returns a ranked list of people based on the demanded set of interest areas.

Based on the discussion on personalization techniques and the depicted graph model in Fig. 3.1, we refine the hubs and authorities approach as follows:

$$A(u; Q) = (1 - \lambda_a)p(u; Q) + \lambda_a \sum_{(u,r) \in E_B} w_{ur}H(r; Q) \tag{3.4}$$

$$H(r; Q) = (1 - \lambda_h)p(r; Q) + \lambda_h \sum_{(v,r) \in E_B} w_{vr}A(v; Q) \tag{3.5}$$

Using this model, the authority of user u is computed with respect to a specific context that is given by query Q. A central role in this model plays the personalization vector $p(u;Q)$. Notice, also the importance ["hubness" depicted by $H(r;Q)$] of a repository r is computed with respect to Q. However, since $A(u;Q)$ is personalized by assigning non-uniform weights to $p(u;Q)$ also $H(r;Q)$ will be influenced by $p(u;Q)$. Thus, with $\lambda_h = 1$ we define $H(r;Q)$ as:

$$H(r;Q) = \sum_{(v,r)\in E_B} w_{vr} A(v;Q) \tag{3.6}$$

We substitute $H(r;Q)$ in Eq. (3.4) and have:

$$A(u;Q) = (1-\lambda)p(u;Q) + \lambda \sum_{(u,r)\in E_B} \sum_{(v,r)\in E_B} w_{ur} w_{vr} A(v;Q) \tag{3.7}$$

The next step is to decompose the query Q as follows:

$$A(u;Q) = (1-\lambda) \sum_{q\in Q} w_q p(u;q) + \lambda$$
$$\sum_{(u,r)\in E_B} \sum_{(v,r)\in E_B} \sum_{q\in Q} w_{ur} w_{vr} w_q A(v;q) \tag{3.8}$$

The weight w_q is associated with a particular keyword q with $w_q = \frac{1}{|Q|}$ for uniform weights and $\sum_q w_q = 1$. In the next steps we apply some of the ideas of the PageRank linearity theorem to our proposed ranking model in Eq. (3.8). This is possible because Eq. (3.8) has a PageRank-like structure. The PageRank linearity theorem has been originally introduced by [34] to create topic-sensitive importance scores for Web-pages, but has not been applied in follow recommendations.

For constant λ we show that authority can be computed as $A(u;Q) = \sum_{q\in Q} w_q A(u;q)$. We restructure Eq. (3.8) to first iterate over each $q \in Q$:

$$A(u;Q) = \sum_{q\in Q} w_q (1-\lambda)p(u;q) + \sum_{q\in Q} w_q \lambda$$
$$\sum_{(u,r)\in E_B} \sum_{(v,r)\in E_B} w_{ur} w_{vr} A(v;q) \tag{3.9}$$

The final step is shown in Eq. (3.10):

$$A(u;Q) = \sum_{q\in Q} w_q \Big[(1-\lambda)p(u;q) + \lambda$$
$$\sum_{(u,r)\in E_B} \sum_{(v,r)\in E_B} w_{ur} w_{vr} A(v;q)\Big] \tag{3.10}$$
$$= \sum_{q\in Q} w_q \big[A(u;q)\big]$$

The model as depicted by Eq. (3.10) brings the following important benefits:

- Ranking scores for individual interest areas can be precomputed and saved in a database. This is typically done periodically in an offline manner.
- At query time, the precomputed ranking scores are retrieved and aggregated to a composite score. This step is performed online at low computational cost.

The proposed model computes user authority with respect to a certain interest area by using personalization techniques. This is a different mechanism than (1) performing matching of users based on interest areas and then (2) computing authority without using personalization techniques. In our approach, authority of users is computed by considering *all repositories* the user has performed actions and giving preference to those repositories by using interest specific personalization weights. This mechanism better captures community-wide reputation by not only computing authority based on a small portion of the user-repository graph but instead using the entire user-repository graph. Thus, our approach follows a PageRank-like model.

In addition, our proposed model is important due to the computational complexity and the inability to compute personalized rankings in an online manner only. For large social networks and collaborative platforms such as the GitHub, the computation of ranking scores may take a long time depending on available hardware and resources.

3.4.3 Weights and Personalization Metrics

The edge weight w_{ur}, which is not personalized or "context" dependent (cf. also Fig. 3.1 and related discussions), is calculated as follows:

$$w_{ur} = \frac{\sum_{t\in T} f(u,r,t)}{\sum_{z\in R(u)} \sum_{t\in T} f(u,z,t)} \tag{3.11}$$

The set T denotes event types (i.e., the type of user action). The set $R(u)$ depicts all repositories u is connected to in G_B (u's neighbors). The function $f(u,r,t)$ retrieves the event count of user u in repository r of event type t.

As mentioned before, personalization is done for individual interest areas and ranking scores are precomputed offline. Assume α is an interest area and $p(u;\alpha)$ is the corresponding personalization vector for users. Generally, we assign to $p(u;\alpha)$ an interest area specific weight and non-interest area specific weight. For brevity, let $R(u;\alpha) = \text{match}(u,\alpha)$ be the set of u's repositories matching the interest area α. Equation (3.12) defines the personalization vector $p(u;\alpha)$:

$$p(u;\alpha) = w_1 \sum_{r\in R(u;\alpha)} w_{ur} + w_2 k_{in}(u;G) / \sum_{v\in V_U} k_{in}(v;G) \tag{3.12}$$

with $w_1 + w_2 = 1$. The first part $\sum_{r \in R(u;\alpha)} w_{ur}$ assigns interest area specific weights to $p(u;\alpha)$. If a given user performs many actions in interest area related repositories (i.e., $R(u;\alpha)$) then also the area specific weight will be higher. The interest area related weight will be more important than the other part and thus $w_1 > w_2$. The second part $k_{in}(u;G)/\sum_{v \in V_U} k_{in}(v;G)$ is the non-interest area specific weight where $k_{in}(u;G)$ depicts u's indegree (follower count) in the follower graph G. This weight increases the likelihood that users with high reputation in terms of follower count are ranked higher than users that are not followed. However, because higher importance is given to the interest area specific weight, we ensure that users have primarily relevant experience and not only many followers.

3.5 Data Collection

3.5.1 GitHub Community

At a high level, GitHub provides information regarding *entities* and *events* through a REST-based API.[2] Information regarding entities are, for example, user details (followers, followings, personal user details) or details regarding repositories (e.g., watchers). Thus, entity information captures the current state of users, repositories, and artifacts. GitHub also provides access to events that describe the dynamic view and state changes. Events are generally generated by user actions. Technical details regarding the REST API interface as well as the event structure can be found in [21] and online at [44]. The GitHub archive [45] offers access to GitHub events. Events can be downloaded in JSON format and inserted into a database for query processing.

We collected GitHub data using the following two methods:

- We retrieved the entire follower/following graph using the GitHub API. The graph was obtained in December 2012.
- We further downloaded all GitHub events between March 11, 2012 and December 12, 2012 from the GitHub archive [45] amounting for roughly 10 months of event data. When performing this research, event data was available starting from March 11, 2012.

Figure 3.2 gives an overview of the different event types[3] captured in the 10-months time frame. The total number of events captured is 36,094,501. The number of distinct users associated with these events is 1,052,916. The total number

---

[2]http://developer.github.com/v3/.

[3]**Event Types**: 1 *PushEvent*, 2 *CreateEvent*, 3 *WatchEvent*, 4 *IssueCommentEvent*, 5 *IssuesEvent*, 6 *ForkEvent*, 7 *PullRequestEvent*, 8 *GistEvent*, 9 *FollowEvent*, 10 *GollumEvent*, 11 *CommitCommentEvent*, 12 *PullRequestReviewCommentEvent*, 13 *MemberEvent*, 14 *DeleteEvent*, 15 *DownloadEvent*, 16 *PublicEvent*, 17 *ForkApplyEvent*.

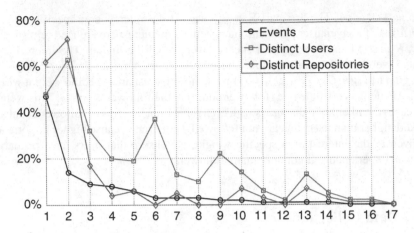

Fig. 3.2 Basic event statistics by type

of distinct software repositories is 2,334,921. The most common event is the *PushEvent* amounting for 46 % of the overall number of events. A PushEvent is a commit on a repository. Second is the *CreateEvent* with 14 % that is triggered when an object ("repository", "branch", or "tag") was created. *WatchEvent, IssueCommentEvent, IssuesEvent* amount for 9 %, 8 %, and 5 % respectively. Detailed descriptions regarding the semantics of event types can be found online [44]. As expected, most events result from actual contributions such as repository commits and the creation of objects.

Table 3.1 provides a list of most popular programming languages based on the count of users watching repositories on GitHub. Language popularity is certainly more biased towards the Linux community where the base technology Git emerged from. However, this bias does not have impact on our follow recommendation approach. On GitHub, javascript is the most popular language with most watchers. It has more than twice as many watchers then the second ranked language.

To understand repository popularity, Fig. 3.3 shows repository watch distributions. Note, all scatter plots in this chapter have logarithmic scale on both the x-axis and y-axis. The actual distribution of the number of repositories shown against the number of users is depicted by Fig. 3.3a. The number of languages watched by how many users is shown by Fig. 3.3b. The depicted distribution in Fig. 3.3a follows a power law with $N(k) \sim k^{-1.97}$ with 132,446 users watching one repository and one user watching 28,080 repositories. In total 406,376 repositories are watched by 327,222 users. Figure 3.3b the number of users over the number of watched languages. The basic characteristics are that 152,577 users watch exactly one language type and the maximum number of languages is 73 watched by one actor.

Table 3.1 Top-20 watched
languages

Rank	Language	Watchers ↓
1	javascript	1,004,137
2	ruby	401,694
3	python	273,070
4	objective-c	243,615
5	php	212,604
6	java	170,359
7	c	128,394
8	c++	92,470
9	shell	63,258
10	c#	49,934
11	coffeescript	44226
12	viml	42595
13	scala	26948
14	perl	26757
15	clojure	24525
16	go	20771
17	emacs lisp	13948
18	erlang	13834
19	actionscript	13339
20	haskell	11378

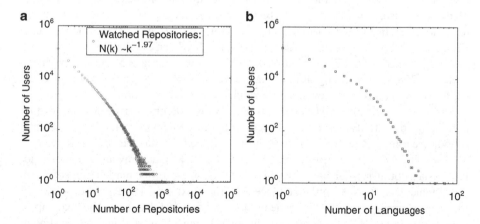

Fig. 3.3 Watched repository and number of languages. (**a**) Watched repositories. (**b**) Number
languages

3.5.2 User Activity by Location

The next step in our analysis is to map user activity to geographic location. User
activity is observed through the number of events (regardless of event type). We
extract location information from events and group events by location and repository

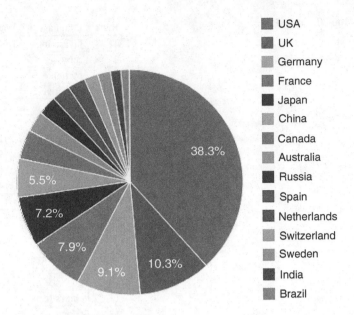

Fig. 3.4 Number of events by country

language. We obtained triples of location, programming language, and frequency of event occurrence. Based on the resulting triples, we perform manual filtering to remove invalid information. Incomplete information (e.g., missing country) has been corrected where possible. The resulting data has been overlaid on a world map and color-coded by frequency. This will be later shown in Fig. 3.5. In total, we have collected 758 location, programming language, and frequency of event occurrence triples based on 5,615,375 events with valid location information.[4]

First of all, we give a summary of activity by country. Figure 3.4 shows the percentage of activity by country based on the aforementioned number of events (about five million events). As shown in Fig. 3.4, the USA has the largest share with 38.3 % followed by three European countries—the UK, Germany and France—and two Asian countries—Japan and China. India, an emerging market for software development, still ranks only at place 14. The next step is to provide detailed location information by showing a heatmap based on GitHub events. The heatmap is weighted by event count (best viewed in color online). The resulting map is shown by Fig. 3.5.

Most user activity clearly happens in the USA and in Europe (see Fig. 3.5). Within these regions, activities span also the largest geographic location with events originating from many different cities. In the USA, most events originate from cities

---

[4]Location information is valid if the location can be mapped correctly to a geographic place. GitHub users may provide false location information which we are unable to control or validate.

Fig. 3.5 Mapping user activities to location

located on either the east- or the west-coast. Events in Europe are scattered among many different cities. In Russia, for example, events almost exclusively originate from Moscow.

3.5.3 Programming Languages

In the following we compare the actor interest across programming languages. The question we attempt to answer is whether actors prefer to watch a pair of programming languages (e.g., java and c#) and what are the similarities between languages. This is expressed as the overlap of watchers. Suppose the set of users U_{l_1} watches language l_1 and the set of users U_{l_2} watches language l_2, than the overlap similarity of watchers by language is calculated as:

$$sim_{watchers}(U_{l_1}, U_{l_2}) = \frac{|U_{l_1} \cap U_{l_2}|}{|U_{l_1}|} \qquad (3.13)$$

Figure 3.6 shows a matrix plot using a color grid to depict watcher similarities. We selected the top-20 languages to compare the joint interest of watchers across programming languages. Notice, the overlap similarity metric used in this context [cf. Eq. (3.13)] is asymmetric and thus the matrix in Fig. 3.6 is also asymmetric. Both axis show programming languages sorted by popularity: left-right (x-axis) and top-down (y-axis).

Javascript-based repositories have the largest number of watchers and thus most other languages share a high degree of similarity in terms of common watchers with javascript (see first column). Scala and clojure have high similarities with java (see column 6), which is not surprising due to their close relationships to java at

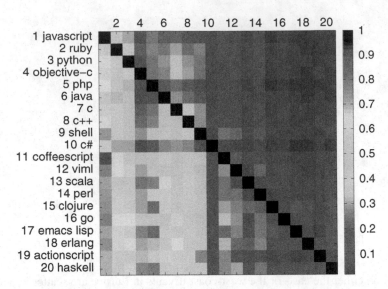

Fig. 3.6 Overlap similarities of watchers per language

a technical level (both languages target the java virtual machine as an execution platform). The languages go, erlang, and haskell have high similarities with the c language (see column 7). Other languages have surprisingly low similarities with the language c# and vice versa (see column and row 10).

Figure 3.7 shows the hierarchical clustering plot which was created based on the overlap similarities of watchers per language. The "correlation" metric was used to calculate the pairwise distance between pairs of objects. The agglomerative hierarchical cluster tree was created through the average linkage method to measure the distance between clusters. Cluster quality was evaluated using the cophenetic correlation coefficient to compare quality against different metrics (e.g., "correlation" metric versus "city block metric"). The value of the coefficient using the "correlation" metric and average linkage was 8.13, which denotes a suitable configuration when compared with other configurations (i.e. metrics).

The observation in Fig. 3.7 are consistent with the previous discussions and present a good clustering of languages. A low distance presents a high similarity based on high watcher overlaps. The language with the lowest similarity (c#) is shown at the bottom of the figure.

3.5.4 Follower Graph and Reciprocity

Follower Graph Not only user-repository relations are used in our recommendation model but also follower relations. Therefore, we analyze GitHub's follower graph. The follower graph counts 1,105,150 users and 1,898,034 following relations

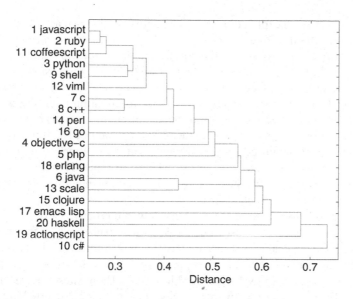

Fig. 3.7 Hierarchical clustering of languages

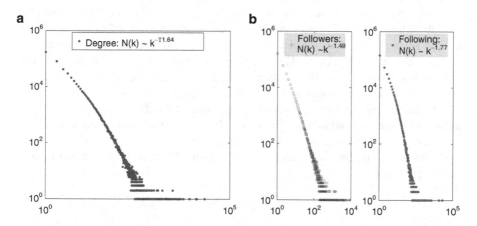

Fig. 3.8 Follower graph degree distributions. (**a**) Degree. (**b**) Followers/following

(edges). Figure 3.8 plots the degree, the indegree (how many followers a given user has) as well as the outdegree (how many users a given user is following) distributions of the graph G.

One can observe a typical power-law distribution with degree $N(k) \sim k^{-1.64}$, indegree (Followers) with $N(k) \sim k^{-1.49}$ and outdegree (Followings) with $N(k) \sim k^{-1.77}$. Nearly 70 % of users (767,975) have no followers (zero indegree) and again about 70 % of users (769,283) do not follow any other user (zero outdegree). Power-law degree distributions are an indicator that the GitHub follower graph exhibits similar structural characteristics as other social networks.

Reciprocity is the concept in directed social networks that two actors are mutually connected to each other (see for example [22]). In a network such as the follower network of GitHub or also Twitter this means that the reciprocal relation between u and v is given if u follows v and v follows u. In the following discussions we will call the reciprocal neighbors of a given user "reciprocal friends". We analyze reciprocity in the context of our work because reciprocity information could be a valuable source for the follow recommendation model. In addition, analyzing reciprocity helps understanding the primary use of the following feature. The following feature could be used to maintain personal relations or on the contrary to follow popular community members ("stars" or "gurus" which are typically hubs in the social network). Depending on the primary use, the recommendation model could be adjusted to the needs of the platform.

The GitHub follower network has very low reciprocity. Only 13 % of user pairs have a reciprocal relationship and 87 % of user pairs have an uni-directional relationship. To compare GitHub's reciprocity with other social network sites: Flickr [26] has a reciprocity of 68 % and a social networking site for personal communications operated by Yahoo! [28] has 84 %. These sites have much higher reciprocity. Twitter has also a low reciprocity of only 22 % as reported in [7]. Also on GitHub, 86 % of users are not followed by their followings.

These observations with regards to reciprocity yield the conclusion that the GitHub follower graph is to a large degree comparable to an information service such as Twitter rather than a typical social (friend) network. A possible explanation for this behavior is that on GitHub many users follow other popular users to get updates regarding their coding activity instead of maintaining personal relations. Our recommendation model is well suited for such a social network because it is based on the concept of authority and reputation. If reciprocity would be higher, other recommendation techniques such as the triadic closure model utilizing (existing) personal relations may be considered as well.

Based on the presented GitHub data, we performed various ranking experiments, which are shown next.

3.6 Evaluation

We perform ranking experiments to analyze the quality of the follow recommendation approach. We use the previously developed recommendation model as detailed Sect. 3.4. The goal of our evaluation is to understand the quality and impact of the personalization techniques. This is done by comparing the top-k results of differently personalized rankings and their results. Since all data and user profiles are public available, we perform ranking and check the GitHub homepages and activities of the top-ranked users. The results can also be easily verified by the reader.

For λ we use a value of $\lambda = 0.9$ for all experiments. A usual value for λ is within the range $0.8 \leq \lambda \leq 0.9$. The metric weights w_1 and w_2 are set to $w_1 = 0.9$ and

Table 3.2 Top-10 follow
recommendations without
personalization

Rank	User	k_{out}	Followers
1	mojombo	9	8,570
2	torvalds	2	8,377
3	defunkt	21	8,373
4	schacon	19	5,748
5	paulirish	61	5,694
6	jeresig	3	4,925
7	pjhyett	1	4,509
8	ryanb	139	4,081
9	android	86	3,889
10	visionmedia	299	3,725

$w_2 = 0.1$; thereby giving higher importance to expertise area specific personalization weights. The weights and personalization are calculated based on the frequency of the various GitHub events.

3.6.1 Recommendations Without Personalization

The first results are calculated without personalization (i.e., the interest area-specific metrics and personalization weights are all zero). The top-10 results are shown by Table 3.2. The actual GitHub user profile can be found online using the link: https:// github.com/User.

The top-10 list in Table 3.2 gives a list of distinguished software engineers and community contributors. Users at rank 1 (mojombo), 3 (defunkt), 4 (schacon), and 7 (pjhyett) are GitHub staff members and, not surprisingly, rank among the top. The Linux father torvalds ranks at position 2. At positions 5 (paulirish) ranks a front-end developer and Google Chrome developer relations engineer. At 6 (jeresig) ranks the creator of the jQuery javascript library. At 8 (ryanb) ranks a producer of ruby on rails screencasts and at 9 (android) ranks the "android" user associated with the android framework, kernel, and system core. At 10 (visionmedia) ranks a javascript contributor. The third column in Table 3.2 shows the outdegree k_{out} in G_B (i.e. the number of repositories the user is involved in). There is a correspondence between rank and number of followers. It becomes already apparent that the top-10 ranked users in Table 3.2 may have very diverse expertise areas with regards to programming languages. We observe a mixture of "javascript", "c", and "java" experts.

3.6.2 Personalized Recommendations

Thus, as the next step we take full advantage of personalization and perform ranking for the interest area "javascript". Table 3.3 shows the top-10 ranking results. The second column shows users and repositories. For each user we provide the top-5 repositories by number of user contributions.

The repository language (if available) is provided in parenthesis next to the repository name. The actual number of repository actions has been omitted for privacy reasons. Furthermore, k_{out} depicts the outdegree of a user in G_B (the number of repositories a user has working on), k_{in} depicts the repository indegree (the number of contributors), and the column *Followers* showing the numbers of followers.

By looking at the results in Table 3.3, two aspects become immediately apparent:

1. The top-10 list is populated by users that contribute to popular javascript repositories.
2. The list is no longer correlated in a strict order by the number of followers.

Both observations demonstrate the intended behavior of our ranking algorithm. Not only high reputation in terms of number of followers is a predominant factor, but also the number of contributions to relevant repositories boosts user authority. By doing so, users gain expertise in a given interest area and are thus ranked at high positions by our ranking algorithm.

Looking at the actual ranked list, the users ranked at 1 (fat) and 2 (caniszczyk) are employees at Twitter and provide contributions to the most popular javascript repository. Ranked at position 3 (mdo) is a designer who is employed at GitHub. What all users have in common is that they are actively engaged in javascript development and are also mostly followed by many people. However, this is not a requirement to be ranked at a top position (for example, see user ranked at 2 who has 91 followers).

To evaluate other interest areas and combinations of languages, we have performed ranking for the keywords {"*java*", "*scale*", "*closure*"} (see Table 3.4) and also {"*objective-c*", "*c*"} (see Table 3.5). Due to space constraints, we provide only the top-3 ranked users as follow recommendation.

As clearly visible in Table 3.4, the top-rankings change in favor of the demanded interest areas (i.e., java and related languages). All top-3 ranked actively contribute to java-related repositories. The first ranked user is also popular in terms of number of followers. Second ranked is the Apache Software Foundation, which is clearly a top-authority in java software development. The GitHub user apache can be regarded as a *proxy* for the real persons behind the account. The third user contributes to similar or the same repositories (e.g., "storm") as the first ranked users. Storm is a distributed realtime computing system in some features similar to Hadoop. Although present in the query, scala repositories are not listed among the top-3 (top-listed) repositories. Finally, we show the results for {*'objective − c', 'c'*} in Table 3.5.

Table 3.3 Top-10 follow recommendations for "javascript"

Rank	User and top-5 repositories	k_{out} / k_{in}	Followers
1	fat	30	1,597
	/twitter/bootstrap (js)	4	
	/twitter/bower (js)	8	
	/maker/ratchet js	3	
	/twitter/hogan.js (js)	3	
	/twitter/recess (js)	2	
2	caniszczyk	34	91
	/twitter/bootstrap (js)	4	
	/twitter/bower (js)	8	
	/twitter/ambrose (js)	4	
	/twitter/twitter.github.com (js)	1	
	/twitter/bootstrap-server (js)	3	
3	mdo	6	879
	/twitter/bootstrap (js)	4	
	/twitter/bootstrap-server (js)	3	
	/mdo/github-buttons	1	
	/mdo/code-guide	1	
	/mdo/sublime-snippets	1	
4	paulirish	61	5,694
	/h5bp/html5-boilerplate (js)	8	
	/twitter/bower (js)	8	
	/yeoman/yeoman js)	7	
	/h5bp/mobile-boilerplate (js)	4	
	/paulirish/infinite-scroll (js)	3	
5	addyosmani	46	3,110
	/addyosmani/todomvc (js)	3	
	/twitter/bower (js)	8	
	/addyosmani/backbone-fundamentals (js)	1	
	/addyosmani/jquery-ui-bootstrap (js)	3	
	/yeoman/yeoman (js)	7	
6	visionmedia	299	3,725
	/visionmedia/express (js)	1	
	/visionmedia/jade (js)	1	
	/visionmedia/mocha (js)	2	
	/senchalabs/connect (js)	1	
	/component/component (js)	2	
7	jzaefferer	53	302
	/jquery/jquery (js)	10	
	/jquery/jquery-mobile (js)	14	
	/jquery/jquery-ui (js)	9	
	/jzaefferer/jquery-validation (js)	2	
	/jquery/qunit (js)	5	

(continued)

Table 3.3 (continued)

Rank	User and top-5 repositories	k_{out} / k_{in}	Followers
8	wycats	43	3,259
	/jquery/jquery (js)	10	
	/rails/rails (ruby)	21	
	/emberjs/ember.js (ruby)	7	
	/wycats/handlebars.js (js)	3	
	/tildeio/rsvp.js (js)	3	
9	scottgonzalez	47	261
	/jquery/jquery (js)	10	
	/jquery/jquery-mobile (js)	14	
	/jquery/jquery-ui (js)	9	
	/jquery/qunit (js)	5	
	/jquery/plugins.jquery.com (js)	3	
10	mbostock	17	1,477
	/mbostock/d3 (js)	1	
	/square/cubism (js)	1	
	/square/crossfilter (js)	1	
	/square/cube (js)	2	
	/d3/d3-plugins (js)	6	

Table 3.4 Top-3 follow recommendations for "java", "scala", and "clojure"

Rank	User and top-5 repositories	k_{out} / k_{in}	Followers
1	nathanmarz	23	857
	/nathanmarz/storm (java)	2	
	/nathanmarz/cascalog (clojure)	2	
	/nathanmarz/storm-starter (java)	1	
	/nathanmarz/storm-contrib (java)	9	
	/nathanmarz/storm-deploy (clojure)	2	
2	apache	242	0
	/apache/cassandra (java)	1	
	/apache/incubator-cordova-android (java)	1	
	/apache/mahout (java)	1	
	/apache/lucene-solr (java)	1	
	/apache/hadoop-common (java)	1	
3	jasonjckn	9	21
	/nathanmarz/storm (java)	2	
	/nathanmarz/storm-contrib (java)	9	
	/nathanmarz/storm-deploy (clojure)	2	
	/nathanmarz/storm-mesos (java)	2	
	/jasonjckn/storm	1	

Table 3.5 Top-3 follow recommendations for "objective-c" and "c"

Rank	User and top repositories	k_{out} / k_{in}	Followers
1	soffes	45	1,100
	/nothingmagical/cheddar-ios (objective-c)	2	
	/soffes/sstoolkit (objective-c)	1	
	/soffes/sspulltorefresh (objective-c)	1	
	/soffes/sskeychain (objective-c)	1	
	/soffes/ssziparchive (c)	1	
2	torvalds	2	8,377
	/torvalds/linux (c)	1	
	/torvalds/subsurface (c)	2	
3	php-pulls	23	2
	/php/php-src (c)	2	
	/php/systems (c)	2	
	/php/pecl-networking-mqseries (c)	2	
	/php/web-php (php)	2	
	/php/php-gtk-src (c++)	1	

Here we see also a similar behavior as previously. Personalization results in rankings of users who are experienced in the demanded interest areas. Top ranked is a popular objective-c developer (according to his online user profile) who has also many followers. Second ranked is torvalds who has undoubtedly high expertise in "c". Third is also a proxy-account php-pulls. Similar to apache, the account is a proxy for the real people (experts) behind it.

To summarize the results of our follow recommendation model, users within the top-k ranking results are very active in the context of the specified interest areas by contributing to popular software repositories. This is the main goal of our authority-based recommendation approach. Thus, we conclude that our model is well suited for personalized, context-sensitive follow recommendations.

3.7 Conclusions

Online software development has taken a new path where social networking features can be used to discover users and repositories. These features help users to stay up-to-date regarding new development efforts and community activities. Here we proposed a novel follow recommendation approach that is based on the concept of user authority. Instead of simply matching users by static skill profiles, we proposed a network-centric approach taking a user's community engagement as well as social metrics into account. We have systematically derived a mathematically sound model to measure user authority based on activity (e.g., repository commits) and community reputation (follower degree).

The presented concepts have been evaluated using a real world dataset. To date, GitHub is the most popular platform offering collaboration features, Wikis, development related tools (e.g., issue tracking) and social networking features such as following. We have obtained a GitHub-based dataset including the follower graph and relevant user actions. Based on the dataset, we have performed a number of experiments to test the proposed recommendation approach. Results show that our personalized follow recommendation approach delivers better recommendations than non-personalized recommendations.

In our future work, we will study more fine grained repository actions for follower recommendations. This includes analyzing the detailed location of changes in the source code as well as details regarding criticality of bugs fixes.

References

1. Github.com, Github website (last access June 2015). URL www.github.com
2. G. Madey, V. Freeh, R. Tynan, The open source software development phenomenon: An analysis based on social network theory, in: Americas conf. on Information Systems (AMCIS2002), 2002, pp. 1806–1813.
3. F. M. Fisher, R. B. Mancke, J. W. McKie, IBM and the U.S. data processing industry : an economic history, Praeger, New York, N.Y., U.S.A. :, 1983.
4. A. W. Kosner, Github is the next big social network, powered by what you do, not who you know (July 2012). URL http://onforb.es/PX02oJ
5. F. O. Bjørnson, T. Dingsøyr, Knowledge management in software engineering: A systematic review of studied concepts, findings and research methods used, Inf. Softw. Technol. 50 (11) (2008) 1055–1068.
6. R. Conradi, T. Dybå, An empirical study on the utility of formal routines to transfer knowledge and experience, SIGSOFT Softw. Eng. Notes 26 (5) (2001) 268–276.   doi: 10.1145/503271.503246. URL http://doi.acm.org/10.1145/503271.503246
7. H. Kwak, C. Lee, H. Park, S. Moon, What is twitter, a social network or a news media?, in: Proceedings of the 19th international conference on World wide web, WWW '10, ACM, New York, NY, USA, 2010, pp. 591–600.
8. J. Weng, E.-P. Lim, J. Jiang, Q. He, Twitterrank: finding topic-sensitive influential twitterers, in: Proceedings of the third ACM international conference on Web search and data mining, WSDM '10, ACM, New York, NY, USA, 2010, pp. 261–270.
9. P. Gupta, A. Goel, J. Lin, A. Sharma, D. Wang, R. Zadeh, Wtf: the who to follow service at twitter, in: Proceedings of the 22nd international conference on World Wide Web, WWW '13, 2013, pp. 505–514.
10. M. J. Brzozowski, D. M. Romero, Who should i follow? recommending people in directed social networks, Tech. rep., HP Labs (2011).
11. J. Xu, Y. Gao, S. Christley, G. Madey, A topological analysis of the open source software development community, in: System Sciences, 2005. HICSS '05. Proceedings of the 38th Annual Hawaii International Conference on, 2005, p. 198a.
12. J. Paulson, G. Succi, A. Eberlein, An empirical study of open-source and closed-source software products, Software Engineering, IEEE Transactions on 30 (4) (2004) 246–256.
13. G. Valetto, M. Helander, K. Ehrlich, S. Chulani, M. Wegman, C. Williams, Using software repositories to investigate socio-technical congruence in development projects, MSR '07, IEEE Computer Society, Washington, DC, USA, 2007, pp. 25–.

14. M. Pinzger, N. Nagappan, B. Murphy, Can developer-module networks predict failures?, in: Proceedings of the 16th ACM SIGSOFT International Symposium on Foundations of software engineering, SIGSOFT '08/FSE-16, ACM, New York, NY, USA, 2008, pp. 2–12.

15. T. Zimmermann, N. Nagappan, Predicting defects using network analysis on dependency graphs, in: Proceedings of the 30th international conference on Software engineering, ICSE '08, ACM, New York, NY, USA, 2008, pp. 531–540.

16. P. Bhattacharya, M. Iliofotou, I. Neamtiu, M. Faloutsos, Graph-based analysis and prediction for software evolution, in: M. Glinz, G. C. Murphy, M. Pezzè (Eds.), ICSE, IEEE, 2012, pp. 419–429.

17. B. Heller, E. Marschner, E. Rosenfeld, J. Heer, Visualizing collaboration and influence in the open-source software community, in: Proceedings of the 8th Working Conference on Mining Software Repositories, MSR '11, ACM, New York, NY, USA, 2011, pp. 223–226.

18. A. Majumder, S. Datta, K. Naidu, Capacitated team formation problem on social networks, in: Proceedings of the 18th ACM SIGKDD international conference on Knowledge discovery and data mining, KDD '12, ACM, New York, NY, USA, 2012, pp. 1005–1013.

19. T. Lappas, K. Liu, E. Terzi, Finding a team of experts in social networks, in: Proceedings of the 15th ACM SIGKDD international conference on Knowledge discovery and data mining, KDD '09, ACM, New York, NY, USA, 2009, pp. 467–476.

20. A. Anagnostopoulos, L. Becchetti, C. Castillo, A. Gionis, S. Leonardi, Online team formation in social networks, in: Proceedings of the 21st international conference on World Wide Web, WWW '12, ACM, New York, NY, USA, 2012, pp. 839–848.

21. G. Gousios, D. Spinellis, Ghtorrent: Github's data from a firehose, in: M. Lanza, M. D. Penta, T. Xi (Eds.), MSR, IEEE, 2012, pp. 12–21.

22. S. Wasserman, K. Faust, Social Network Analysis: Methods and Applications, Cambridge University Press, Cambridge, 1994.

23. D. J. Watts, S. H. Strogatz, Collective dynamics of 'small-world' networks, Nature 393 (6684) (1998) 440–442.

24. M. E. J. Newman, J. Park, Why social networks are different from other types of networks, Phys. Rev. E 68 (2003) 036122.

25. J. Leskovec, E. Horvitz, Planetary-scale views on a large instant-messaging network, in: Proceedings of the 17th international conference on World Wide Web, WWW '08, ACM, New York, NY, USA, 2008, pp. 915–924.

26. M. Cha, A. Mislove, K. P. Gummadi, A measurement-driven analysis of information propagation in the flickr social network, in: Proceedings of the 18th international conference on World wide web, WWW '09, ACM, New York, NY, USA, 2009, pp. 721–730. doi:10.1145/1526709.1526806.

27. D. Schall, Expertise ranking using activity and contextual link measures, Data Knowl. Eng. 71 (1) (2012) 92–113.

28. R. Kumar, J. Novak, A. Tomkins, Structure and evolution of online social networks, in: Proceedings of the 12th ACM SIGKDD international conference on Knowledge discovery and data mining, KDD '06, ACM, New York, NY, USA, 2006, pp. 611–617. doi:10.1145/1150402.1150476.

29. D. Nevo, I. Benbasat, Y. Wand, Who knows what? (Oct. 2009). URL http://sloanreview.mit.edu/executive-adviser/2009-4/5147/who-knows-what/

30. A. Mockus, J. D. Herbsleb, Expertise browser: a quantitative approach to identifying expertise, in: Proceedings of the 24th International Conference on Software Engineering, ICSE '02, ACM, New York, NY, USA, 2002, pp. 503–512. doi:10.1145/581339.581401. URL http://doi.acm.org/10.1145/581339.581401

31. T. Fritz, J. Ou, G. C. Murphy, E. Murphy-Hill, A degree-of-knowledge model to capture source code familiarity, in: Proceedings of the 32nd ACM/IEEE International Conference on Software Engineering - Volume 1, ICSE '10, ACM, New York, NY, USA, 2010, pp. 385–394. doi:10.1145/1806799.1806856. URL http://doi.acm.org/10.1145/1806799.1806856

32. J. M. Kleinberg, Authoritative sources in a hyperlinked environment, J. ACM 46 (5) (1999) 604–632.

33. L. Page, S. Brin, R. Motwani, T. Winograd, The pagerank citation ranking: Bringing order to the web, Tech. rep., Stanford Digital Library Technologies Project (1998).
34. T. H. Haveliwala, Topic-sensitive pagerank, in: WWW '02, ACM, New York, NY, USA, 2002, pp. 517–526.
35. G. Jeh, J. Widom, Scaling personalized web search, in: WWW '03, ACM, New York, NY, USA, 2003, pp. 271–279.
36. D. Schall, Measuring contextual partner importance in scientific collaboration networks, Journal of Informetrics 7 (3) (2013) 730–736. doi:http://dx.doi.org/10.1016/j.joi.2013.05.003. URL http://www.sciencedirect.com/science/article/pii/S1751157713000461
37. J. Zhang, M. S. Ackerman, L. Adamic, Expertise networks in online communities: structure and algorithms, in: Proceedings of the 16th international conference on World Wide Web, WWW '07, ACM, New York, NY, USA, 2007, pp. 221–230.
38. L. A. Adamic, J. Zhang, E. Bakshy, M. S. Ackerman, Knowledge sharing and yahoo answers: everyone knows something, in: Proceedings of the 17th international conference on World Wide Web, WWW '08, ACM, New York, NY, USA, 2008, pp. 665–674.
39. D. Schall, Service Oriented Crowdsourcing: Architecture, Protocols and Algorithms, Springer Briefs in Computer Science, Springer New York, New York, NY, USA, 2012.
40. M. McPherson, L. S. Lovin, J. M. Cook, Birds of a feather: Homophily in social networks, Annual Review of Sociology 27 (1) (2001) 415–444.
41. M. S. Granovetter, The strength of weak ties, The American Journal of Sociology 78 (6) (1973) 1360–1380.
42. D. Goldberg, D. Nichols, B. M. Oki, D. Terry, Using collaborative filtering to weave an information tapestry, Commun. ACM 35 (12) (1992) 61–70. doi:10.1145/138859.138867. URL http://doi.acm.org/10.1145/138859.138867
43. A. Y. Ng, A. X. Zheng, M. I. Jordan, Stable algorithms for link analysis, in: Proceedings of the 24th annual international ACM SIGIR conference on Research and development in information retrieval, SIGIR '01, ACM, New York, NY, USA, 2001, pp. 258–266.
44. Github.com, Github event types. URL http://developer.github.com/v3/activity/events/types/
45. I. Grigorik, Github archive (last access June 2015). URL www.githubarchive.org

Chapter 4
Partner Recommendation

Abstract In this chapter we present a novel approach for measuring and combing various criteria for partner importance evaluation in scientific collaboration networks. The presented approach is cost sensitive, aware of temporal and context-based partner authority, and takes structural information with regards to structural holes into account. The applicability of the proposed approach and the effects of parameter selection are extensively studied using real data from the European Union's research program.

4.1 Social Network-Based Collaboration

Scientific collaboration in an international environment takes place among partners such as organizations, universities or research institutes to jointly perform projects. The main motivation for organizations and individual research groups to collaborate is to enable knowledge and resource sharing to effectively perform research projects. Scientific collaboration can be defined as *interaction taking place within a social context among two or more scientists that facilitates the sharing of meaning and completion of tasks with respect to a mutually shared, superordinate goal* [1].

However, the success of research and innovation is based on the right balance between cooperation and competition. Hence, formation of coalitions and consortia is influenced by partner reputation [2], institutional constraints, and mechanism of self-organization [3]. Scientific collaboration can be analyzed at the level of researchers through co-authorship and citation networks [4–6] or at the level of organizations or research institutions [7]. The former has been widely studied by existing research while the latter lacks a principled approach for selecting and aggregating ranking criteria that may be influenced by context. Generally, scientific collaboration and endorsement can be analyzed according to three different methods [8]: (1) qualitative methods such as using a questionnaire-based approach, (2) bibliometric methods including publication and citation counting or co-citation analysis, and (3) complex network methods including network centrality metrics such as PageRank [9] or Hyperlink Induced Topic Search (HITS) [10]. Here we focus on the analysis of scientific collaboration at the organizational or institutional level. We apply complex network methods to automate the analysis of partner importance in scientific collaboration. In this work, importance is a concept that

© Springer International Publishing Switzerland 2015
D. Schall, *Social Network-Based Recommender Systems*,
DOI 10.1007/978-3-319-22735-1_4

is governed by multiple factors including average cost of a partner, temporal trend and context of partner authority, and partner importance with regards to effective size of the partner's social network. Effective size in the context of structural holes and social networks means low redundancy among social contacts thereby yielding control benefits of individuals. Here we apply a similar principle but focus on the organizational level rather than individuals in social networks.

In our previous work [11] we introduced an approach for measuring contextual importance in scientific collaboration networks. In this work, we build upon our previous work [11] but significantly expand the concepts. Here we provide the following novel key contributions:

- We introduce a personalized partner authority model that is able to capture context-dependent and time-aware partner reputation.
- We introduce a model to measure structural importance of organizations embedded in scientific collaboration networks. The idea of our structural importance metric is drawn from the notion of structural holes as established in a sociological research context.
- To support partner selection using multiple-criteria, the factors contributing to a partner importance are aggregated through a systematic approach to a single partner importance ranking score. Here we apply analytic hierarchy process (AHP) to derive the partner importance score.
- We present experimental results by providing a comprehensive study on the influence of different parameters using real data from the EU's Seventh Framework Programme (FP7) for research in Information and Communication Technology (ICT).

This chapter is structured as follows. Section 4.2 gives an overview of related work and literature in the context of network formation and network analysis. In Sect. 4.3 our personalized partner authority model is introduced. Section 4.4 introduces the structural importance model and Sect. 4.5 details the software framework and the analytic hierarchy process to compute the final partner importance scores. In Sect. 4.6 the evaluation results are presented followed by the conclusion and outlook to future work in Sect. 4.7.

4.2 Background in Strategic Formation

We structure this background section into two basic areas: *network formation* in the context of collaborative environments and *network analysis* methods with particular emphasis on authority ranking. From a technique point of view, many approaches found in both network formation and network analysis methods for authority ranking are based on graph theory and algorithms. In this section, we review literature in both areas as they will provide the foundation for our work.

Network Formation The rapid advancement of ICT-enabled infrastructure has fundamentally changed how businesses and companies operate. Global markets and the requirement for rapid innovation demand for alliances between individual companies [12]. It is widely agreed that knowledge of the structure of interaction among individuals or organizations is important for a proper understanding of a number of important questions such as the spread of new ideas and technologies and competitive strategies in dynamic markets [13]. Work by [14] investigated the evolutionary dynamics of network formation by analyzing how organizational units create new linkages for resource exchange. The potential gains from bridging different parts of a network were important in the early work of Granovetter [15] and are central to the notion of structural holes developed by Burt [16, 17]. The theory is based on the hypothesis that individuals can benefit from serving as intermediaries between others who are not directly connected. A formal approach to strategic formation based on advanced game-theoretic broker incentive techniques was presented in [18]. In [19] group formation in social networks is studied.

Network Analysis We propose a model for importance that is based on well-established techniques such as the notion of hubs and authorities [10] and PageRank [9]. PageRank can be personalized [9] to estimate node importance with regards to certain topics [20–22]. After the seminal work of [9] and the far-reaching work of [21], related research (see also [23]) addressed, for example, efficient computation of personalized PageRank [24, 25] and a generalization of personalized PageRank towards bipartite graphs [26]. In [27], the authors proposed time-aware authority ranking by considering temporal properties of scientific publication activity. Our previous work addressed PageRank personalization techniques for expertise ranking in a social network context [28, 29].

In this work, we propose a new framework which utilizes both information from structural holes and authority importance scores to discover valuable collaboration partners. Here we propose a unified HITS/PageRank model that is able to measure network importance at the individual as well as the organizational or institutional level with respect to a certain context. In contrast to existing rankings such as the Shanghai academic ranking,[1] our approach is able to capture importance at a fine grained contextual level. Our approach is able to utilize various additional ranking parameters including desirable partner properties (e.g., high topic-sensitive authority) and low undesirable partner properties (e.g., partner costs). At the core of this framework are link-based algorithms such as HITS and extensions towards personalized, time-aware PageRank, structural metrics to measure the brokerage potential of a given network node, and an analytic hierarchy process (AHP) algorithm [30] to aggregate these metrics into a single ranking score.

---

[1] www.shanghairanking.com.

The proposed model is tested with data from the ICT research projects having received grants under the EU's FP7 program. The data as described in [31] and covers a period from 2007 to 2011.

4.3   Reputation Model

Here we formalize the notion of organization authority as it will be used in our ranking model. Authority is automatically calculated using network analysis techniques. The novelty of the approach is that authority is put *into context* by considering topic information. Well-established models provide the foundational concepts and basis. Specifically, we base our approach upon the model of hubs and authorities as developed by [10].

4.3.1   Basic Definitions

We start with a definition of basic concepts that are used throughout this work. Let us consider a simple collaboration scenario in a scientific community where individual partners (e.g., organizations, research institutes, and universities) collaborate in the context of research projects.

Figure 4.1 shows a set of organizations $\{o_1, o_2, o_3\}$ and a set of research projects $\{p_1, p_2, p_3\}$. Each project is associated with a certain topic that determines the *context* of the performed collaboration (for example, "services" or "internet"). Organizations are involved in projects by having certain roles. Roles include project coordinator and project partner. In addition to the involvement relation, a weighted edge is created from the project to the organization to depict the degree of involvement. For example, o_1 is involved in projects p_1 and p_2 with weights

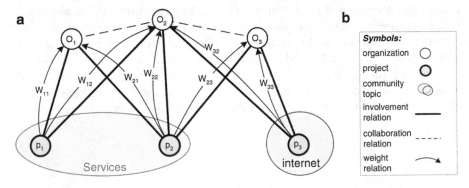

Fig. 4.1 Scientific collaboration environment and definitions. (**a**) Scientific collaboration environment. (**b**) Legend

w_{11} and w_{21} respectively. In our work, the weight will be based on the funding an organization receives in the context of a project. More funding typically means that an organization is able to allocate more (human) resources to the project and thereby performing more work. Finally, based on joint projects performed by organizations we model collaboration relations among them. Since o_1 and o_2 have been involved in the joint projects p_1 and p_2, a collaboration relation between o_1 and o_2 is established as a dashed line. Similarly, o_2 and o_3 have been involved in the joint projects p_2 and p_3 and therefore a collaboration relation between o_2 and o_3 is established. Also, a collaboration relation between o_1 and o_3 exists because they jointly worked on p_2. A collaboration relation is a mutual (undirected) edge.

4.3.2 Hubs and Authorities

Let us apply the notion of hubs and authorities to a collaboration environment as depicted by Fig. 4.1. A project is regarded to be important if the organizations contributing to it are also regarded to be important (e.g., knowledgeable and reputable). In turn, the importance of an organization is based on its involvement in important projects. This is a recursive definition of importance and can be modeled by using the intuitive notion of hubs and authorities as proposed by [10].

$$A(o) = \sum_{(p,o) \in E_P} H(p) \qquad H(p) = \sum_{(p,u) \in E_P} A(u) \qquad (4.1)$$

In the model, an organization o obtains an authority score depicted by $A(o)$ and a project p obtains a hub score denoted by $H(p)$. The drawback of this model is the "stability" of rankings. A ranking algorithm is stable if the algorithm returns similar results upon small disturbances. We follow the *randomized HITS* approach as proposed in [32] and expand the equations in Eq. (4.1) as follows:

$$A(o) = (1 - \lambda_a)\delta_O(o) + \lambda_a \sum_{(p,o) \in E_P} H(p) \qquad (4.2)$$

$$H(p) = (1 - \lambda_h)\delta_P(p) + \lambda_h \sum_{(p,u) \in E_P} A(u) \qquad (4.3)$$

This adjusted model is a natural way of designing a random-walk based algorithm following the HITS model. The randomized HITS approach is, like PageRank, stable to small perturbations [32]. The symbols $\delta_O(o)$ and $\delta_P(p)$ depict personalization vectors that may be assigned uniformly for each node such that $\delta_O(o) = \frac{1}{|V_O|}$ and $\delta_P(p) = \frac{1}{|V_P|}$. Non-uniform personalization vectors result in personalized rankings. The parameters λ_a and λ_h with $0 \leq \lambda \leq 1$ allow for balancing between authority/hub weights and personalization weights. A typical value for λ is 0.85 [9]. Assigning lower values to λ means that higher importance is given to the personalization weights; thereby reducing the "*network effect*" of the ranking algorithm.

4.3.3 Query-Sensitive Personalization

Let us define the query-sensitive authority score:

$$A(o;Q) = (1 - \lambda_a)\delta_O(o;Q) + \lambda_a \sum_{(p,o)\in E_P} w_{po}H(p;Q) \qquad (4.4)$$

Similarly, let us define the query-sensitive hub score:

$$H(p;Q) = (1 - \lambda_h)\delta_P(p;Q) + \lambda_h \sum_{(p,u)\in E_P} w_{pu}A(u;Q) \qquad (4.5)$$

The edge weights w_{po} and w_{pu} are based on the organizations' degree of project involvement. Particularly, the weight w_{po} is based on the funding received by organization o in project p and is calculated as

$$w_{po} = \frac{funding(p,o)}{\sum_{v\in adj(p)} funding(p,v)} \qquad (4.6)$$

where $adj(p)$ depicts the set of nodes adjacent to p (i.e., the set of organizations involved in project p). To compute authority scores using a single equation, which is the desired goal of our approach, we substitute $H(p;Q)$ in Eq. (4.4) by Eq. (4.5) and have:

$$A(o;Q) = (1 - \lambda_a)\delta_O(o;Q) + \lambda_a(1 - \lambda_h) \sum_{(p,o)\in E_P} w_{po}\delta_P(p;Q)$$
$$+ \lambda_a\lambda_h \sum_{(p,o)\in E_P} \sum_{(p,u)\in E} w_{po}w_{pu}A(u;Q) \qquad (4.7)$$

Based on Eq. (4.7), let us define the personalization vector $\delta'_O(o;Q)$ as follows:

$$\delta'_O(o;Q) = \frac{1-\lambda_a}{1-\lambda_h}\delta_O(o;Q) + \lambda_a \sum_{(p,o)\in E_P} w_{po}\delta_P(p;Q) \qquad (4.8)$$

If we use the same parameter values for λ_a and λ_h [due to symmetry of Eqs. (4.4) and (4.5)] such that $\lambda_a = \lambda_h$, Eq. (4.8) simplifies to:

$$\delta'_O(o;Q) = \delta_O(o;Q) + \lambda \sum_{(p,o)\in E_P} w_{po}\delta_P(p;Q) \qquad (4.9)$$

In the following step we rewrite Eq. (4.7) by using the personalization vector $p'(u;Q)$ as defined in Eq. (4.9).

$$A(o;Q) = (1 - \lambda)\delta'_O(o;Q) + \lambda^2 \sum_{(p,o)\in E_P} \sum_{(p,u)\in E_P} w_{po}w_{pu}A(u;Q) \qquad (4.10)$$

As one can see, Eq. (4.10) has a PageRank-like structure. An important concept for personalization based on the PageRank model is the *linearity theorem* as introduced in [21]. The theorem states that for any personalization vectors δ_1, δ_2 and weights w_1, w_2 with $w_1 + w_2 = 1$, the following equality holds:

$$PPV(w_1\delta_1 + w_2\delta_2) = w_1 PPV(\delta_1) + w_2 PPV(\delta_2) \tag{4.11}$$

The linearity theorem states that personalized PageRank vectors *PPV* can be composed as the weighted sum of PageRank vectors. Equation (4.12) shows how to derive the weighted sum of personalized authority ranking scores using Eq. (4.10). The goal is to obtain a structure as depicted by the right part of Eq. (4.11). The weight w_q is associated with a particular keyword q with $w_q = \frac{1}{|Q|}$ for uniform weights and $\sum_q w_q = 1$.

$$
\begin{aligned}
A(o;Q) &= (1-\lambda)\delta'_P(o;Q) + \lambda^2 \sum_{(p,o)\in E_P} \sum_{(p,u)\in E_P} w_{po}w_{pu}A(u;Q) \\
&= (1-\lambda)\sum_{q\in Q} w_q\delta'_P(o;q) + \lambda^2 \sum_{(p,o)\in E_P} \sum_{(p,u)\in E_P} \sum_{q\in Q} w_q w_{po}w_{pu}A(u;q) \\
&= \sum_{q\in Q} w_q(1-\lambda)\delta'_P(o;q) + \sum_{q\in Q} w_q\lambda^2 \sum_{(p,o)\in E_P} \sum_{(p,u)\in E_P} w_{po}w_{pu}A(u;q) \\
&= \sum_{q\in Q} w_q\left[(1-\lambda)\delta'_P(o;q) + \lambda^2 \sum_{(p,o)\in E_P} \sum_{(p,u)\in E_P} w_{po}w_{pu}A(u;q)\right] \\
&= \sum_{q\in Q} w_q\left[A(o;q)\right]
\end{aligned}
\tag{4.12}
$$

As stated before, the benefit of the model is the ability to precompute authority scores for particular topics, save them in a database, and aggregate the precomputed authority scores later at query time. Suppose the set of topics, as extracted by the `Topic Analyzer`, is given as $T = \{T_1, T_2, \ldots, T_n\}$. For each topic authority scores are calculated $A(o;T_1), A(o;T_2), \ldots, A(o;T_n)$ and utilized by the `Authority Aggregator` to compute

$$A(o;Q) = \sum_{q\in Q} w_q A(o;T_q) \tag{4.13}$$

where T_q is the topic matching query keyword q. Next, we describe the time-aware authority model.

4.3.4 Time-Aware Authority

We have extensively discussed the notion of authority and the idea of computing authority scores for individual topics that can be aggregated at query time. Now we turn to the definition of the personalization vectors δ_P and δ_O. Recall, δ_P

holds personalization weights for projects and δ_O holds personalization weights for organizations. For δ_P we use a straightforward model to calculate personalization weights

$$\delta_P(p) = \frac{\text{funding}(p)}{\sum_{proj \in V_P} \text{funding}(proj)} \tag{4.14}$$

where *funding*(p) depicts the monetary funding received by project p. For simplicity, we do not consider the query context Q for the project-based personalization vector.

The next discussion is related to the concept of time-aware and topic-based authority ranking. Thus, we establish metrics to calculate the personalization weights of δ_O. Here topic-based personalization and time-aware weighting is applied. Recall that a topic is identified by a single keyword. Organizations typically perform numerous projects that are related to one or more topic(s). Thus, each organization has a set of topics including topic frequency associated with it. Furthermore, frequencies of topics are counted by year. An example for such data would be ("OrgA", 2011, "services", 5) where "OrgA" is the organization name, 2011 the specific year, "services" the given topic and the number 5 an example of a frequency count. As a first step let us define the weight function $W^T(o, y; T_x)$ that obtains the frequency count of organization o in year y for some topic T_x. The frequency count is based on how many projects related to the given topic the organization has started in the year (i.e., the year when signing the project contract). To establish the notion of positive or negative change in topic specific weights, we define the weight deviation function $W_\Delta^T(o, y; T_x)$ as follows:

$$W_\Delta^T(o, y; T_x) = W^T(o, y; T_x) - \frac{1}{|Y|} \sum_{y' \in Y} W^T(o, y'; T_x) \tag{4.15}$$

Deviation in this context means the weight $W^T(o, y; T_x)$ in year y minus the average weight with regards to topic T_x. Straightforwardly, a positive sign means increasing topic-based weight, a negative sign means decreasing topic-based weight as a result of being below the average, and 0 means no change in topic-based weights (i.e., through constant rate of projects related to topic T_x). This definition is quite simple and captures already a notion of "trend" by analyzing the temporal project history of an organization. The positive/negative sign shows increasing or decreasing trend. However, $W_\Delta^T(o, y; T_x)$ just analyzes the trend with respect to organization o without considering the weights and thus performance of other organizations. Personalization for authority ranking in collaboration networks must be performed by considering weights in relation to all other organizations. For brevity, let us define the set $\alpha = \{W^T(o_1, y; T_x), W^T(o_2, y; T_x), \dots, W^T(o_n, y; T_x)\}$ with $\{o_1, o_2, \dots, o_n\} \in V_O$. Let us define the trend $Tr(o; T_x)$ of organization o with respect to topic T_x as:

$$Tr(o; T_x) = \sum_{y \in Y} w_y \left[\frac{W^T(o, y; T_x)}{max(\alpha)} \times W_\Delta^T(o, y; T_x) \right] \tag{4.16}$$

$Tr(o; T_x)$ is based on the trend for topic T_x over the years $Y = \{y_1, y_2, \ldots, y_n\}$ where y_n is the most recent year, y_{n-1} the previous year and so forth (ordered by recency). The first term within the square brackets measures the topic based weight in relation to the community performance in T_x by dividing $W^T(o, y; T_x)$ by $max(\alpha)$. For the top-performing organizations having the most numbers of projects related to T_x in year y the term becomes 1. The term is multiplied by the organization specific weight deviation function $W_{\Delta}^T(o, y; T_x)$. The weight w_y puts more emphasis on recent years (recency factor) by being calculated as $w_y \in \{\frac{1}{|Y|}, \frac{1}{|Y|-1}, \frac{1}{|Y|-2}, \ldots, 1\}$.

Finally, the personalization vector δ_O needs to be assigned by matching organizations having performed projects related to T_x and trend values Tr need to be mapped to a positive interval. This is done because δ_O represents a probability distribution (for theoretical foundations related to personalized PageRank see, for example, [20]). Let us define the set $\beta = \{Tr(o_1; T_x), Tr(o_2; T_x), \ldots, Tr(o_n; T_x)\}$ with $\{o_1, o_2, \ldots, o_n\} \in V_O$.

$$\delta_O(o; T_x) = \begin{cases} 1 - \frac{max(\beta) - Tr(o; T_x)}{max(\beta) - min(\beta)} & , \text{if } matches(o; T_x) \\ 0 & , \text{otherwise} \end{cases} \qquad (4.17)$$

The function $matches(o; T_x)$ checks if o has performed projects related to T_x and evaluates to true or false. To evaluate a query $Q = \{T_1, T_2\}$ a simple aggregation is performed

$$A(o; \{T_1, T_2\}) = w_1 A(o; T_1) + w_2 A(o; T_2) \qquad (4.18)$$

where $A(o; T_1)$ is personalized for T_1 and $A(o; T_2)$ is personalized for T_2. In other words, both $A(o; T_1)$ and $A(o; T_2)$ hold topic-based and time-aware authority scores for all $o \in V_O$.

4.4   Structural Importance Model

The previous section explained in detail the authority model and ranking approach. Here we turn to the second criteria used in our overall ranking model. We define the notion of structural importance and detail a metric to calculate the importance. The obtained ranking scores for structural importance are used as a second parameter in the AHP-based aggregation [i.e., the AHP parameter $SI(o; Q)$]. By following the notion of structural holes as coined by Burt [16, 17], structural holes are an opportunity to broker the flow of information between people in an organizational or social network. As an example, managers often act as information brokers as they talk to many people in the project.

Structural importance captures the ability of a network node to broker information between its neighbors (in our context organizations). A node can do so if potential "information gaps" (or buffers) arise in the network. A broker can also be seen as a mediator that helps establishing communication between other nodes. A project partner with "brokerage" capabilities is often important in project consortia to help establish and facilitate communication among other partners. As an example, a project consortium may be lead by an academic partner who is in charge of coordinating the project from an administrative and scientific point of view. Typically, exploitation and further use of project results is an important issue in research projects. However, the consortium leader may not be the optimal partner for transferring (or "translating") scientific results to business. Thus, there may be a gap between technical/scientific results and exploitation of results within an industrial context (e.g., implementing novel solutions within an industrial environment). With regards to this example, an organization may act as a broker by mediating communication and transferring the knowledge to an industrial partner within the project.

Thus, structural importance essentially focuses on mediation capabilities of an organization as opposed to expertise/authority. Such mediators help running projects more effectively and efficiently by (a) establishing communication between potentially disconnected network segments that have not communicated before and (b) help making communication more fluid and efficient. To be able to act as a broker, gaps must exist in the network because otherwise a node looses its ability to establish communication. The notion of *redundancy* provides means to express the existence of such gaps. If there is high redundancy in terms of network edges and communication paths in a network, the need to fill structural gaps may be very limited. On the contrary, if a network is highly segmented and only few nodes connect individual segments, the need for brokers and mediation opportunities may be very high.

Let us consider a graph as depicted by Fig. 4.2. Here the graph model G_{OC} is used that consists of organizations and collaboration relations as undirected edges. Each node depicts an organization with $\{a,b,c,o,r,u,v,z\} \subset V_O$. A circle surrounds nodes that belong to a particular expertise area or community identified through A and B. A query may be formulated to match the nodes and edges in either Q^A or

Fig. 4.2 Network structure to illustrate structural importance metric

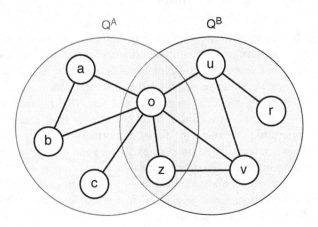

Q^B or both $Q = Q^A \cup Q^B = \{T_A, T_B\}$. An edge $(v, u) \in G_{OC}$ has a weight which is based on the number of performed projects between v and u. The weight is dynamically assigned depending on the query context Q. For example, the weight of the edge $(o, z) \in G_{OC}$ may be different in Q^A and Q^B depending on the joint projects performed by o and z (i.e., if the projects match Q^A or Q^B or both). Suppose $Q = Q^A \cup Q^B$, the node o has the highest number of non-redundant edges in the graph because it connects the node sets $\{a, b, c\}$ and $\{u, v, z\}$ which are only reachable via o. Thus, o has a unique position within the network because o is able to control the information flow between both node sets. Furthermore, only o and z belong to both expertise areas A and B but only o is connected to $\{a, b, c\}$ in A. Let us define $SI(o; Q)$ as

$$SI(o; Q) = \sum_{u \in N(o)} \left[1 - \sum_{v \in N(u)} W_N^Q(o, v; Q) W_M^Q(u, v; Q) \right] \tag{4.19}$$

where $v \notin \{u, o\}$ and $N(o)$ the set of o's neighbors. For $SI(o; Q)$, we follow Burt's measure of the effective size of a node's network [17]. Here the notion of structural holes is not applied to people-based social networks, but to organization collaboration networks (i.e., G_{OC}). Conceptually, the effective size is the number of nodes o is connected to, minus the redundancy in the network.

In contrast to Burt's definition of effective size, we compute structural importance with respect to the query Q. As an example, while o in Fig. 4.2 is structurally important in $Q = Q^A \cup Q^B$ to establish a flow between $\{a, b, c\}$ and $\{u, v, z\}$, o is less significant if only Q^B is considered. Actually, within Q^B u has a unique position because r is only reachable via u.

The weight $W_N^Q(o, v; Q)$ in Eq. (4.19) shows the query-sensitive normalized edge weight between o and v and is calculated as

$$W_N^Q(o, v; Q) = \sum_{q \in Q} \frac{w_{ov}^q}{\sum_{u \in N(o)} w_{ou}^q} \tag{4.20}$$

where w_{ov}^q is the weight associated with $(o, v) \in E_O$ and calculated as the number of joint projects between o and v matching the query keyword q. Furthermore, the weight $W_M^Q(u, v; Q)$ in Eq. (4.19) depicts the query-sensitive marginal edge weight between u and v and is calculated as follows:

$$W_M^Q(u, v; Q) = \sum_{q \in Q} \frac{w_{uv}^q}{max(\{w_{un}^q | \forall n \in N(u)\})} \tag{4.21}$$

The marginal weight of u with neighbor v is the weight w_{uv}^q (also based on the number of matching joint projects between them) divided by u's strongest weight with anyone of its neighbors $N(u)$. If none of the projects match q, the weight is assigned to $w_{uv}^q = 0$.

4.5   Framework and Ranking Algorithm

4.5.1   Software Framework

As already outlined before, our solution approach to support multi-criteria partner
selection in scientific communities utilizes heavily graph-based models. Graph-
based models are widely used in complex- and social-network analysis. Figure 4.3
shows the solution framework as a layered view.

Data Management   The layer underneath the top-layer shows the data management
that is responsible for retrieval of project relevant data, managing the needed graph
structures to perform analysis and ranking, and persistence management of analysis
and ranking results. From the top-layer (Offline analysis) point of view, the data
management can be access via the Data Manipulation Handler in a CRUD
(Create-Read-Update-Delete) manner. The Data Provider offers read access to
graph structures and offline mining and ranking results. The Project Database
contains information such as organizations, projects, project involvements, roles,
funding, project descriptions, date of project contracts, and project duration. Let
us define some basic graph structures that are obtained from information in the
Project Database and then managed in the Graph Database.

The Corporate Policies database can only be queried but not modified by
the framework. Essentially, it contains white and black list information with respect
to preferred or denied partners. The lists are influenced by mid to long-term business
strategy.

Based on projects, organizations, and involvement relations we define two
types of graphs. First, let us define the directed project-organization graph
$G_{PO}(V_P, V_O, E_P)$ that is composed of the set projects V_P and the set of organizations
V_O (V_P and V_O depicting the vertices in the graph) and the project involvement
relations denoted by the edge set E_P where an edge $(p, o) \in E_P$ points from the
project p to organization o. Each edge $(p, o) \in E_P$ has a weight w_{po} associated with it
depending on the funding the organization o receives in project p divided by the total
project funding. This type of graph is being used for *organization authority ranking*.
Let us define the second type of graph as the undirected organization-collaboration
graph $G_{OC}(V_O, E_O)$ consisting of the set of organizations V_O depicting the vertices
in the graph and the set of collaboration relations E_O. Whereas the edges E_P in G_{PO}
are based on project involvement relations pointing from projects to organizations,
the edges E_O in G_{OC} are undirected and connect two organizations. This type of
graph is being used for *structural importance ranking*. Details regarding these two
types of graph structures will be provided in the following.

Offline Analysis   This layer deals with components that are invoked in an offline
manner (e.g., triggered by changes in the Project Database). The Topic
Analyzer extracts relevant topics from project descriptions by filtering stop words
and combining synonyms to single topics. Essentially, each topic is identified by
a single keyword that has a frequency associated with it to identify popularity of

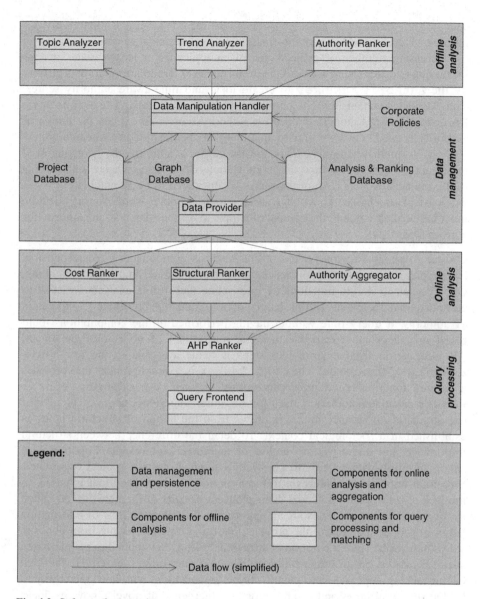

Fig. 4.3 Software framework

topics. Typical topics in the context of ICT research are, for example, "services" and "internet". Sophisticated topic models such as cross-topic relations or hierarchical structures are not within the focus of this work (e.g., see [29] for hierarchical topic models and topic clustering techniques). The Topic Analyzer saves topic information in the Analysis & Ranking Database. If new topics are added, other components such as the Authority Ranker are triggered (see later).

The `Trend Analyzer` calculates trends with regards to organizations' activities in topics. The historical project information is used to calculate a trend in increasing or decreasing number of projects for given topics. In a steady state (neither increasing nor decreasing activity), the trend equals 0. Trend information is utilized by the `Authority Ranker` to create topic and time-aware authority scores. The detailed mechanisms will be discussed in later sections. The `Authority Ranker` calculates numeric values for authority scores. To explain the notion of authority as used in this work, participation of an organization in a research project (i.e., involvement relation) is understood as a carrier of authority. By being involved in certain projects, we assume that organizations develop knowledge with regards to the projects' topic(s). An organization is considered to be an authority if it has extensive or specialized knowledge about a topic. In other words, an organization must have collaborated in the context of a topic to be considered as an authority for a given topic.

Online Analysis Previously, the offline analysis components were responsible for preparing the information needed to perform online analysis and ranking. Our decision to divide functionality into online and offline analysis was due to computational complexity of link-based authority ranking algorithms. Computation of authority at query time without having performed offline computation would result in unacceptable response times at magnitudes of hours or even longer. All information from the previous steps is made available in the `Analysis & Ranking Database`. The `Cost Ranker` is a simple ranker that provides a scoring function based on organizations' average costs. The `Structural Ranker` calculates numeric values for structural importance scores. The idea of our structural importance metric is drawn from the notion of structural holes as established in a sociological context. To detail the difference between structural importance and authority, the notion of authority captures the importance of an organization with regards to knowledge drawn from past project experience. Structural importance captures a different notion of importance that is based on the lack of information flow and connectedness of parts of the network. As stated by Burt [16], structural holes are an opportunity to broker the flow of information between people and control the projects that bring together people from opposite sides of the hole. Here the notion of structural holes is not applied to people-based social networks, but to organization collaboration networks (i.e., G_{OC}). The goal of our ranking approach is to identify those organizations who have the ability to bridge structural holes and to allow for the emergence of novel innovative ideas through brokerage of information. The `Authority Aggregator` combines the authority results of offline computed authority scores.

Query Processing Suppose a coordinator attempts to establish a new consortium and thus wants to find collaboration partners who are able to join the consortium. Often, previous collaborators are known from first hand collaboration experience but in today's vibrant and fast-paced research environment it is also useful to see the current community standing of known collaboration partners and to discover potential new collaborators. The coordinator is able to specify a keyword-based query $Q = \{q_1, q_2, \ldots, q_n\}$ (using the `Query Frontend`) with the goal of finding matching organizations that are ranked according to a set of criteria (i.e., cost, structural importance and authority). The idea of our ranking approach is to compute ranking scores with respect to certain *areas of expertise*. The demanded areas of expertise are specified via the query Q and matched with topics. Each query keyword q_n corresponds to a desired area of expertise. A query returns a ranked list of organizations based on the demanded set of expertise areas. The `AHP Ranker` is used to create a composite ranking score $S(o; Q)$ of organization o. The score $S(o; Q)$ is given as

$$S(o; Q) = AHP(A(o; Q), SI(o; Q), C(o)) \qquad (4.22)$$

where $A(o; Q)$ is the organization's authority score and $SI(o; Q)$ the structural importance score with respect to the query Q, and $C(o)$ the cost score. The following section shows the calculation steps.

4.5.2   Ranking Algorithm

Here we discuss the computation of the final ranking score. Recall that the composite ranking score $S(o; Q)$ of organization o is obtained through the function $AHP(A(o; Q), SI(o; Q), C(o))$. Previously we have defined the authority $A(o; Q)$ and the structural importance $SI(o; Q)$. Cost $C(o)$ is calculated as the average funding organization o receives:

$$C(o) = \frac{1}{num\_projects(o)} \sum_{(p,o) \in E_P} funding(p, o) \qquad (4.23)$$

The final aggregation and computation of a composite ranking score is done using the AHP algorithm. AHP is a technique for making complex decisions in a structured way. AHP has been successfully applied in a number of fields including transportation [33], maintenance and configurations [34], and service quality assessment [33]. The theoretical background will not be covered in this work since AHP is a well explored technique. We refer the reader to [30] for details regarding AHP as a decision making technique.

Algorithm 4 shows the main steps at a high level. The input of the algorithm is given as the query Q and the organization-collaboration graph G_{OC}. The graph G_{OC} is used to compute the structural importance scores in an online manner. Next four essential steps are performed: (1) create map with criteria input scores, (2) setup AHP, (3) perform AHP ranking, and (4) assign final AHP ranking scores to output map.

First, the ranking criteria scores are obtained as described in the previous sections (authority Sect. 4.3 and structural importance Sect. 4.4 respectively). These include

Algorithm 4 Multi-criteria ranking algorithm

Input: The query Q and the undirected organization-collaboration graph G_{OC}.
Compute:

1. Create map for org with individual scores. For each organization $o \in V_O$ do:

   - $A(o) \leftarrow au\_score(o, Q)$ // authority
   - $SI(o) \leftarrow si\_score(o, G_{OC}, Q)$ // struct. imp.
   - $C(o) \leftarrow avg\_cost(o)$ // cost
   - Add to map $(o, \{A(o), SI(o), C(o)\})$

2. Setup AHP attributes weights and desirability.

   - Auth. attributes ("*authority*", $\{w_{au}, +1\}$)
   - Struct. attributes ("*structure*", $\{w_{si}, +1\}$)
   - Cost attributes ("*cost*", $\{w_{cost}, -1\}$)

3. Perform AHP ranking using output from previous steps.

   - Compute the vector of criteria weights.
   - Compute the matrix of organizations scores.
   - Rank the organizations.

4. Assign final AHP ranking scores to map S. For each organization $o \in V_O$ do:

   - $S(o) \leftarrow ahp\_score(o)$ // final score

Output: Ranked organizations based on query Q and according to composite AHP ranking score.

authority, structural importance and cost. Using a map, each criteria score is associated with an organization. The map generated in this step is passed as an argument to the AHP ranking in step 3.

Second, AHP attributes are setup by assigning the weights w_{au}, w_{si}, w_{cost} to each criteria with $[\sum_w w] = 1$. In addition, the desirability attribute is assigned to denote if a certain criteria is desired or not. In particular, authority and structural importance should be high (desirability $= +1$) to obtain a better AHP ranking score whereas cost should be low to obtain a better ranking score (desirability $= -1$).

Third, AHP ranking is performed by using the previously setup attributes and the output map of step 1. The step 3 of Algorithm 4 is decomposed into the following steps:

- Compute the vector of criteria weights: In this step rating of the relative priority of the criteria is done by assigning a weight value to the more important criteria. The weight values are taken from the previous step of the algorithm (step 2). The weight assignment is done through a pairwise comparison of the criteria. After that, the resulting weights are normalized and the average is computed for each criteria.
- Compute the matrix of organizations scores: Here the score for each organization is determined by computing how well organization o meets some criteria Y. Afterwards, the organizations' scores are normalized and averaged.

- Rank the organizations: In a final step the organizations' scores are combined with the criterion weights to produce an overall score for each organization. The extent to which the organizations satisfy the criteria is weighted according to the relative importance of the criteria. The final score is simply computed as a weighted sum.

Note, in our case criteria are contrasting by demanding that organizations should have high authority but low cost. In general, the organization that is recommended for selection (top-ranked in final output S) is not necessarily the one which optimizes each single criterion, but rather the organization which achieves the most suitable *trade-off among the different criteria*. This behavior makes AHP a very flexible and powerful tool for multi-criteria partner selection.

Forth, the AHP scores are saved in a final score map S. Organizations are ranked in descending order by ranking score.

4.6   Evaluation

Here the evaluation of the proposed concepts and model is presented. We have selected a dataset of a scientific collaboration environment to test the concepts.

4.6.1   Description of Dataset

The data is based on ICT research projects having received grants under the EU's Seventh Framework Programme (FP7). The data as described in detail in [31] and covers a period from 2007 to 2011. Research projects have multiple partners and an organization can be the partner of multiple projects. To date, the FP7 ICT program has allocated funding to 1,469 projects for a total Union funding of 4,979,301,152 Euro. This results in 14,781 participations by 4,718 distinct legal entities. The Fig. 4.4 (Source: European Commission[2]) shows a clear geographic concentration, across and within the member states.

Our evaluation is performed as follows. First we select the top-20 organizations (ranked by degree and given in Table 4.2) and compute metrics for those 20 organizations with regards to popular topics. This evaluation is called top-k rank evaluation and is presented in Sect. 4.6.2. Second we compute cross topic ranking statistics such as overlap similarity and Kendall's τ rank difference. This evaluation is presented in Sect. 4.6.3 along with the definition of relevant ranking metrics.

Table 4.1 gives an overview of popular (project) topics extracted from project information (see [31] for details). Frequency is measured by counting appearance

---

[2]http://observatory.euroris-net.eu/euroris/files/download/261-316.

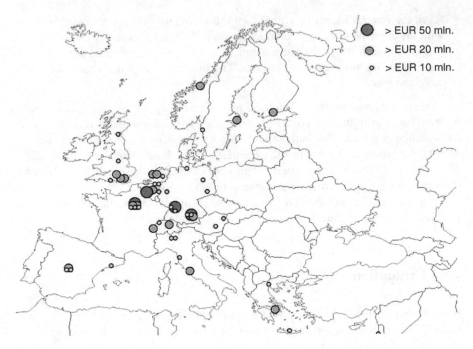

Fig. 4.4 Location of organisations the top 50 for EC funding from FP7-ICT

Table 4.1 Popular project
topics and frequencies

Topic	Frequency
Systems	4126
Internet	2729
Networks	1771
Services	1247
Software	1224
Health	1115
Embedded	1054
Transport	890
Efficiency	849
Energy	849

of the topic string within project names and project short descriptions of each project
partner involvement record (association of organization to project including received
funding). In total, we extracted 170 topics after performing some automatic and
manual processing of the data. Table 4.1 shows the top-10 topics with the highest
frequencies among the 170 topics.

Table 4.2 Top-20 organizations ranked by degree

PNr	Name	Cost	Degree	Struct. rank score
1	Fraunhofer-Gesellschaft zur Foerderung der Angewandten Forschung E.V	524,515	272	516
2	Centre National De La Recherche Scientifique	159,983	153	175
3	Commissariat A L Energie Atomique Et Aux Energies Alternatives	235,911	137	201
4	Ecole Polytechnique Federale De Lausanne	290,827	97	146
5	Consiglio Nazionale Delle Ricerche	455,650	96	154
6	Valtion Teknillinen Tutkimuskeskus	293,240	95	190
7	Institut National De Recherche En Informatique Et En Automatique	799,995	94	127
8	Interuniversitair Micro-Electronica Centrum Vzw	964,195	90	140
9	Eidgenoessische Technische Hochschule Zurich	389,544	90	89
10	Telefonica Investigacion Y Desarrollo Sa	636,818	76	131
11	Katholieke Universiteit Leuven	711,085	69	98
12	SAP AG	1221,665	68	168
13	Universidad Politecnica De Madrid	331,315	65	136
14	Atos Origin Sociedad Anonima Espanola	628,296	62	215
15	Imperial College Of Science, Technology And Medicine	286,930	61	56
16	Politecnico Di Milano	531,975	61	98
17	Kungliga Tekniska Hoegskolan	415,684	59	85
18	Technische Universiteit Delft	622,286	58	88
19	Karlsruher Institut Fuer Technologie	250,599	56	76
20	Technische Universitaet Wien	287,363	55	87

4.6.2   Top-k Rank Evaluation

Table 4.2 shows the top-20 organizations ranked by their degree in G_{PO} (project-organization graph). The first column (**PNr** column) is a unique key associated with an organization and used throughout this section to identify a top-20 organization. The second column (**Name** column) shows the organizations' legal name. The third column (**Cost** column) shows the average organization cost using Eq. (4.23). The organization indegree (**Degree** column) is analog to the project count as projects $p \in V_P$ point to organizations $o \in V_O$. The degree-based rank will be used as a baseline ranking. This baseline results will be compared with AHP-based rankings. We have selected the degree-based rank as a baseline algorithm to show the impact of personalization based on topic information and time-aware authority ranking. Notice, the degree-based rank has no topic bias. In addition, the degree-based rank was selected because it already captures some notion of importance with regards to organization reputation. The last column (**Structural Rank Score** column) shows the structural rank score [using Eq. (4.19)] over all topics in Table 4.1. A higher score is better.

It is noticeable that the organization 14 has a particular high structural rank score in relation to its degree-based rank position. Organization 1 has an exceptional high structural rank score but has also the most projects within the ICT framework program. Notice, however, the degree is calculated using G_{PO} and the structural rank using G_{OC}.

To compare AHP results with the rankings in Tables 4.2, 4.3 and 4.4 list detailed metrics for selected topics. We have selected eight out of the ten topics from Table 4.1 due to space reasons. Each metric is computed for each topic in Tables 4.3 and 4.4 respectively. Using G_{PO}, the degree **D** is based on matching projects only. Projects are matched against the given topic as depicted in the headings of Tables 4.3 and 4.4. The authority **A** is calculated for respective topics. The position **P** is the rank position index as obtained by the AHP rank using Eq. (4.22).

AHP is setup with the weights 0.4 for authority, 0.2 for the structural importance rank and 0.4 for cost. Thus, authority and cost are given slightly higher weights than structural importance. We regard authority as highly desirable but at the same time cost should be kept at an acceptable level. After that structural importance is also a desirable property but not equally important as the other criteria. However, since our approach is flexible weights can be adjusted as demanded.

The position change **Ch** is computed between degree-based ranking positions and authority based ranking positions in the following manner

$$Ch(o) = pos(A(o; T_x)) - pos(degree\_rank(o)) \qquad (4.24)$$

where $pos()$ retrieves the position index by ranking score. This lets us show how rankings are influenced by authority. Finally, the trend **Tr** is computed by using the Eq. (4.16). As state before, the sign has the following meaning:

$$Tr = \begin{cases} \text{positive sign} & \text{, if trend is increasing} \\ \text{negative sign} & \text{, if trend is decreasing} \\ 0 & \text{, otherwise} \end{cases}$$

To show the relationship between two metrics, at the bottom of Tables 4.3 and 4.4 we show the correlation coefficient among various metrics. As usual, the correlation coefficient takes a value between $[-1, 1]$, with 1 or -1 indicating perfect correlation. A positive correlation shows a positive association between the variables. Thus, increasing values of one variable correspond to increasing values of the other variable. On the other hand, negative correlation indicates a negative association between the variables. Thus, increasing values of one variable correspond to decreasing values of the other variable. A correlation value close to 0 indicates no association between the variables.

Table 4.3 Top-20 list of organizations: topics include "networks", "systems", "software", and "services"

PNr	Networks					Systems					Software					Services				
	D	A	P	Ch	Tr	D	A	P	Ch	Tr	D	A	P	Ch	Tr	D	A	P	Ch	Tr
1	29	0.11	1	2	1.08	70	0.86	1	0	7.97	7	0.15	2	3	0.23	11	0.04	5	28	−0.03
2	10	0.03	31	677	−0.13	39	0.00	16	196	−0.03	3	0.04	12	27	0.00	3	0.04	15	35	0.00
3	20	0.05	12	25	0.19	44	0.10	5	4	0.79	1	0.04	7	20	0.00	1	0.04	10	28	0.00
4	5	0.03	30	665	−0.05	26	0.01	19	87	0.01	1	0.04	16	27	0.00	2	0.04	17	34	0.00
5	4	0.05	16	27	0.17	19	0.15	3	−1	1.44	6	0.06	10	6	0.04	7	0.04	11	13	0.06
6	12	0.07	6	5	0.56	20	0.05	8	15	0.39	1	0.04	9	19	0.00	7	0.02	25	630	−0.18
7	16	0.09	5	−2	0.85	25	0.08	10	5	0.73	7	0.04	18	44	−0.01	9	0.04	22	618	−0.02
8	6	0.04	27	61	0.01	27	0.14	6	−3	1.21	0	0.01	42	357	0.00	1	0.04	19	25	0.00
9	7	0.04	32	31	0.10	33	0.09	12	−1	0.87	1	0.04	29	18	0.00	1	0.04	34	27	0.00
10	38	0.00	681	671	−0.55	6	0.01	26	67	0.06	11	0.01	37	353	−0.07	12	0.04	18	16	0.03
11	3	0.03	57	661	−0.05	23	0.05	18	8	0.48	2	0.04	25	25	0.00	2	0.04	32	32	0.00
12	7	0.07	7	2	0.51	6	0.01	15	56	0.06	17	0.22	3	−10	0.43	18	0.17	2	−10	1.25
13	6	0.06	13	6	0.38	12	0.01	22	67	0.06	10	0.03	20	349	−0.03	11	0.00	636	625	−0.34
14	9	0.10	3	−10	1.06	5	0.02	9	38	0.12	7	0.05	5	1	0.03	11	0.11	3	−9	0.72
15	0	0.00	689	667	0.00	19	0.04	42	13	0.38	1	0.04	57	28	0.00	2	0.04	59	31	0.00
16	3	0.04	44	88	0.00	22	0.07	14	−3	0.66	5	0.05	19	−2	0.03	5	0.04	30	16	0.02
17	9	0.04	42	48	0.04	19	0.09	13	−8	0.83	1	0.04	34	21	0.00	1	0.04	37	25	0.00
18	3	0.03	60	642	−0.03	25	0.15	7	−15	1.48	0	0.00	101	348	0.00	2	0.04	36	26	0.00
19	2	0.04	65	84	0.00	17	0.02	49	26	0.18	1	0.04	45	33	0.00	1	0.04	47	33	0.00
20	3	0.04	52	96	0.00	17	0.02	40	30	0.15	3	0.04	32	6	0.01	5	0.06	13	−9	0.20
D	1.0	0.2	0.3	0.0	0.1	1.0	0.8	−0.4	−0.1	0.8	1.0	0.6	−0.4	0.1	0.6	1.0	0.5	0.2	0.3	0.5
A		1.0	−0.6	−0.7	1.0		1.0	−0.4	−0.3	1.0		1.0	−0.5	−0.5	1.0		1.0	−0.4	−0.4	1.0
P			1.0	0.6	−0.5			1.0	0.2	−0.4			1.0	0.5	−0.4			1.0	0.5	−0.3
Ch				1.0	−0.6				1.0	−0.3				1.0	0.0				1.0	−0.4

Table 4.4 Top-20 list of organizations: topics include "health", "embedded", "internet", and "energy"

PNr	Health					Embedded					Internet					Energy				
	D	A	P	Ch	Tr	D	A	P	Ch	Tr	D	A	P	Ch	Tr	D	A	P	Ch	Tr
1	12	0.05	3	22	0.07	11	0.12	2	10	0.51	47	0.04	3	11	0.47	17	0.43	1	0	0.56
2	7	0.06	10	9	0.26	12	0.23	4	1	1.22	14	0.03	24	1040	-0.21	0	0.00	25	609	0.00
3	3	0.06	11	14	0.19	15	0.27	3	-1	1.38	23	0.03	8	33	0.12	3	0.05	10	12	0.01
4	5	0.07	9	5	0.30	8	0.04	23	41	0.04	8	0.03	21	114	-0.02	4	0.03	14	24	0.00
5	8	0.06	14	11	0.21	1	0.03	25	49	0.00	11	0.03	15	36	0.13	3	0.04	12	14	0.01
6	4	0.04	20	37	0.00	3	0.03	19	46	0.01	15	0.05	6	5	0.67	14	0.08	7	3	0.07
7	6	0.03	37	699	-0.05	12	0.21	5	-2	1.07	27	0.05	7	0	0.83	2	0.03	19	24	0.00
8	0	0.00	714	706	0.00	10	0.05	20	25	0.11	6	0.03	20	65	0.01	0	0.00	49	602	0.00
9	9	0.02	710	703	-0.16	11	0.02	59	506	-0.05	9	0.03	33	57	0.05	1	0.03	32	20	0.00
10	1	0.04	28	34	0.00	2	0.03	33	48	0.00	54	0.00	1044	1034	-1.41	2	0.03	17	20	0.00
11	5	0.04	25	15	0.07	2	0.03	41	48	0.00	5	0.03	29	52	0.07	2	0.03	26	23	0.00
12	0	0.00	713	703	0.00	5	0.04	17	28	0.08	28	0.09	2	-10	2.60	6	0.21	2	-10	0.27
13	12	0.06	13	0	0.25	5	0.07	14	7	0.26	18	0.03	23	61	0.04	2	0.03	16	22	0.00
14	4	0.04	19	38	0.00	0	0.00	24	510	0.00	20	0.12	1	-13	4.42	5	0.04	9	6	0.01
15	9	0.11	4	-11	0.67	5	0.06	29	7	0.23	1	0.03	77	95	0.00	3	0.06	18	-2	0.05
16	4	0.04	34	32	0.00	9	0.04	32	23	0.10	8	0.03	27	33	0.12	3	0.03	24	16	0.00
17	0	0.00	720	699	0.00	9	0.05	28	14	0.15	10	0.03	39	52	0.05	0	0.00	163	597	0.00
18	3	0.04	31	12	0.04	10	0.15	10	-9	0.72	4	0.03	42	88	0.00	0	0.00	137	595	0.00
19	3	0.04	44	40	0.00	6	0.11	12	-5	0.54	3	0.03	60	120	0.00	4	0.05	21	-5	0.02
20	4	0.06	15	-10	0.27	11	0.02	60	494	-0.04	8	0.03	52	1004	-0.04	2	0.03	36	35	0.00
D	1.0	0.6	-0.4	-0.3	0.4	1.0	0.7	-0.2	-0.1	0.6	1.0	0.1	0.6	0.2	0.1	1.0	0.8	-0.4	-0.4	0.8
A		1.0	-0.8	-0.8	0.9		1.0	-0.7	-0.4	1.0		1.0	-0.4	-0.4	1.0		1.0	-0.4	-0.4	0.8
P			1.0	0.9	-0.4			1.0	0.7	-0.7			1.0	0.6	-0.4			1.0	0.7	-0.3
Ch				1.0	-0.5				1.0	-0.4				1.0	-0.4				1.0	-0.2

Table 4.3 shows the results for the topics "networks", "systems", "software", and "services". The organization PNr 1 has been ranked by AHP at position 1 in "networks", position 1 in "systems", position 2 in "software", and position 5 in "services". With regards to "services", a negative trend is shown for PNr 1 and thus the position has dropped in this topic. In the other topics, positive trend can be observed and thus the ranking position was mostly preserved. With regards to the topic "systems", a very good trend of 7.97 can be observed and highest authority score of 0.86 within the table. As one can see, by applying our approach, much more fine-grained ranking can be performed by considering topic information.

With regards to correlation, **A** always correlates perfectly with **Tr** because time-aware authority takes trend through personalization in account. **D** shows good correlation with **Tr** in the topic "systems". This is a result of the broad scope of "systems" and the high frequency of the topic within projects (see also Table 4.1).

Table 4.4 shows the results for the topics "health", "embedded", "internet", and "energy". The organization PNr 1 was only ranked in "energy" at position 1 but not for the other topics. One exceptional high change in the ranking position can be seen for organization PNr 10 in "internet" which ranks by AHP at 1044. PNr 10 had some substantial amounts of projects with regards to "internet" in the past (54 matching projects as indicated by **D**) but the trend is highly negative (**Tr** is -1.41, which is the lowest in the table) and time-aware **A** is 0.00. Thus, we believe that negative trend and limited recent activity in the context "internet" justifies a change in the rank position.

With regards to correlation, **A** only correlates perfectly in "embedded" and "internet" but not for the other topics (although a high correlation is still achieved). **D** shows good correlation with **Tr** in the topic "energy". Like in Table 4.3, **A** shows good correlation with **P**. Indeed, authority is part of AHP's ranking criteria so a correlation can be expected. Recall, higher authority yields better positions. Thus, negative correlation means increasing values of authority correspond to decreasing values of the rank position (lower position value is better).

Based on the data in Tables 4.3 and 4.4, average values of degree, position, and change are depicted in Fig. 4.5 and average values of authority and trend are shown in Fig. 4.6. Average values are based on the metric values of the top-20 list of organizations. Again, the baseline algorithm for ranking is the degree-based rank.

The topic "networks" has the highest average value with regards to change. Thus, AHP rankings based on topic information have significant impact on the ranking position of organizations and a lot of changes are observed within the top-20 list. Also the topics "health" and "internet" yield high changes on average. However, only "health" yields also high average values with regards to position. This means that organizations ranked within top-20 positions by the degree-based rank would be ranked at much higher positions by AHP in the "health" topic. As mentioned before, since "systems" is a very broad topic also the positions by AHP are quite similar when compared with the degree-based rank (the lowest average value as depicted by Fig. 4.5). The average degree does not significantly change across topics. Generally, topic based personalization has the effect that significant changes of rank position can be expected.

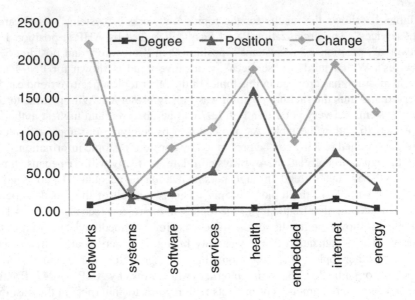

Fig. 4.5 Average degree, position, and change

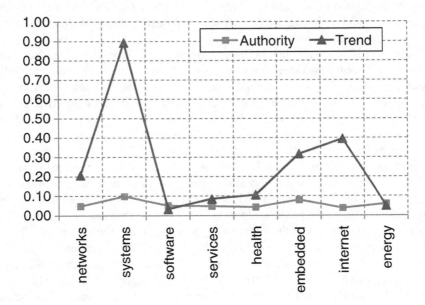

Fig. 4.6 Average authority score and trend

Next, Fig. 4.6 shows the average values for authority and trend. The topic "systems" shows the highest average authority and the highest average trend. This observation is also consistent with the previous discussion. The topic "software" shows the lowest trend and also a low average value for authority. In general, deviations in authority across topics is very high.

To summarize the main observations in this section:

- Our proposed model enables more fine-grained ranking by considering topic information.
- Authority correlates to a high degree with trend because time-aware authority takes trend through personalization into account.
- Generally, topic based personalization has the effect that significant changes of rank position can be observed.
- Topics that play a role in many projects (having a broad scope) correlate better with degree-based ranking. Thus, no significant changes through personalization can be expected.
- As a consequence of the previous observation, by focusing on narrow and more specialized topics organizations with fewer projects are able to build up authority and are thereby ranked at better positions in those topics.

4.6.3 Statistical Comparison

Here a statistical comparison of ranking techniques is performed. In the previous section, a top-20 list of organizations was selected (as ranked by the organizations' degree) and evaluated by using different metrics. In this section we use a set overlap and distance based ranking metric to compare the AHP based results with non-personalized rankings including the degree-based rank, a funding based rank, and the structural rank.

The funding based rank uses the total amount of funding received by an organization to perform ranking (the higher the total funding the better the rank). The structural importance rank is used in isolation of AHP and compared with the regular AHP using the criteria authority, structural importance, and cost. After that a cross topic comparison is performed by using AHP and authority based rankings and AHP-based rankings personalized for different topics. AHP is setup with the weights 0.4 for authority, 0.2 for the structural importance rank and 0.4 for cost.

To systematically compare results of two ranking algorithms, let us define two standard ranking metrics.

OSim@k To measure similarity of top-k sets, let us define overlap similarity as follows:

$$\text{OSim@}k = \frac{O_{k1} \cap O_{k2}}{k} \tag{4.25}$$

OSim@k defines the overlap similarity of the top-k sets O_k ranked by algorithm 1 and algorithm 2. Each set consists of organizations such that $O_k \subset V_O$. The first algorithm is always AHP, which has been parameterized using the same weights as defined previously.

Kendall's τ The next ranking metric used in this work is the well-known Kendall's τ metric (for example, see [28]):

$$\text{Kendall's } \tau = \frac{2(num\_concordant - num\_disconcordant)}{|V_O|(|V_O| - 1)} \tag{4.26}$$

Consider the pair of nodes o, u. The pair is concordant if two rankings agree on the order and disconcordant if both rankings disagree on the order. Denote the number of these pairs by *num_concordant* and *num_disconcordant* respectively. The total number of pairs is given as $\frac{|V_O|(|V_O|-1)}{2}$. Kendall's τ is defined between the interval $\tau \in [-1, 1]$. Kendall's τ helps analyzing if two ranking algorithms are rank similar. If τ equals 1, there are no cases where the pair o, u is ranked in a different order.

Table 4.5 shows the comparison results of AHP-based rankings (for the top-10 topics in Table 4.1) and the degree-based, funding-based, and structural importance rank. The highest values for OSim and Kendall's τ are depicted as bold-face numbers. The topic "systems" clearly shows the highest overlap with the other (non-topic based) rankings. OSim@10, OSim@20, and OSim@50 show the highest overlap in each topic. This observation is again in line with the previous discussion. Previously "systems" showed the highest average authority and the highest average trend within the top-20 list of organizations. The structural importance rank shows the highest overlap of 0.70 in the top-10 segment (depicted as OSim@10). However, an higher agreement in the rank order as measured through Kendall's τ is given in the topic "software". Kendall's τ is calculated by using the whole list of ranked organizations. Whereas the highest overlap of AHP-based rankings with the degree-based, funding-based, and structural importance rank is given in "systems", higher agreement in terms of Kendall's τ is given in "software".

Figure 4.7 shows the comparison results of AHP-based rankings (again for the top-10 topics in Table 4.1) and the authority-based rankings. Here, for each topic ranking is performed using AHP as defined in Eq. (4.22) and authority as defined in Eq. (4.18). The results are then compared using OSim and Kendall's τ. Further details are provided in Table 4.6. The first set of rows (1–10) shows OSim@10, the second set of rows (11–20) depicts OSim@20, the third set of rows (21–30) depicts OSim@50, and the forth set of rows (31–40) shows Kendall's τ.

The values below the matrix diagonal (from top-left to bottom right corner) are all set to 0 because of symmetry. For example, overlap similarity OSim for the topics "networks" and "systems" yields the same results as "systems" and "networks". At the diagonal values comparison of AHP and authority rankings for the same topic was performed. Thus, high overlap and agreement with regards to OSim and Kendall's τ, respectively, can be observed.

Figure 4.8 shows the average values of OSim@10, OSim@20, OSim@50, and Kendall's τ for each topic. With regards to OSim@10, "health" yields the lowest average overlap similarity. The topics "efficiency" and "energy" have the highest overlap similarities in the top-10 segment.

Table 4.5 OSim and Kendall's τ for comparison of AHP with degree, funding, and structural rank

		Networks	Systems	Software	Services	Transport	Efficiency	Health	Embedded	Internet	Energy
Degree	OSim@10	0.30	0.60	0.40	0.20	0.30	0.30	0.30	0.40	0.40	0.30
	OSim@20	0.40	0.75	0.55	0.50	0.40	0.55	0.50	0.50	0.40	0.55
	OSim@50	0.56	0.86	0.72	0.74	0.70	0.78	0.66	0.68	0.70	0.78
	Kendall's τ	0.44	0.42	0.46	0.45	0.44	0.41	0.41	0.45	0.43	0.41
Funding	OSim@10	0.30	0.50	0.50	0.30	0.30	0.40	0.30	0.30	0.40	0.40
	OSim@20	0.40	0.70	0.50	0.55	0.40	0.55	0.40	0.40	0.60	0.55
	OSim@50	0.56	0.84	0.76	0.72	0.70	0.76	0.74	0.66	0.70	0.76
	Kendall's τ	0.63	0.54	0.65	0.64	0.60	0.58	0.61	0.61	0.62	0.58
Structural	OSim@10	0.40	0.70	0.60	0.40	0.40	0.60	0.30	0.30	0.50	0.60
	OSim@20	0.40	0.70	0.65	0.55	0.45	0.60	0.40	0.45	0.60	0.60
	OSim@50	0.56	0.86	0.80	0.82	0.80	0.84	0.66	0.72	0.68	0.84
	Kendall's τ	0.48	0.47	0.51	0.49	0.49	0.46	0.46	0.50	0.46	0.46

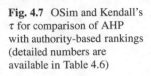

Fig. 4.7 OSim and Kendall's
τ for comparison of AHP
with authority-based rankings
(detailed numbers are
available in Table 4.6)

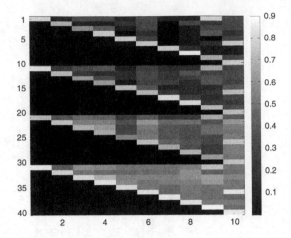

Figure 4.9 shows the comparison results of AHP-based rankings for the top-10 topics in Table 4.1 across topics. This comparison shows how ranking results change by considering different topics. Further details are provided in Table 4.7.

Rows are segmented in the same manner as already described previously. Values at the matrix diagonal (from top-left to bottom right corner) are all 1. The values below the matrix diagonal are all set to 0 for the before mentioned reason.

Figure 4.10 shows the average values of OSim@10, OSim@20, OSim@50, and Kendall's τ for each topic. In OSim@10 the topic "health" results in the lowest average overlap similarity followed by the topic "embedded", which has also low overlap similarity. In general, higher average values of OSim@10, OSim@20, OSim@50 as well as Kendall's τ can be observed when compared with the previous discussion. Higher values are the result of the same ranking technique being used (AHP-based rankings) and results being compared across topics. Before the AHP-based rankings were compared with authority, which is only one of the ranking criteria being used in AHP.

Overall, the overlap in the top-10 segment is on average 42 %, in the top-20 segment 49 %, and in the top-50 segment 69 %. This means that around six out of ten organizations in the top-10 would be ranked differently across topics. Thus, personalization using topic information has a strong impact on ranking results. For the topic "health", for example, it has the largest impact with average OSim@10 being 23 %. We observe changes of more than 49 % in some topics by looking at OSim@10.

4.7 Conclusions

This work introduced various metrics for importance ranking in scientific collaboration environments. We proposed a novel topic-sensitive authority model that is based on well-establish ranking techniques. We systematically derived a unified

Table 4.6 OSim and Kendall's τ for comparison of AHP with authority-based rankings

		Networks	Systems	Software	Services	Transport	Efficiency	Health	Embedded	Internet	Energy
OSim@10	Networks	0.80	0.10	0.30	0.30	0.10	0.30	0.10	0.10	0.80	0.30
	Systems	0.00	0.70	0.10	0.10	0.20	0.30	0.00	0.30	0.20	0.30
	Software	0.00	0.00	0.60	0.40	0.10	0.30	0.00	0.10	0.30	0.30
	Services	0.00	0.00	0.00	0.80	0.00	0.20	0.00	0.10	0.40	0.20
	Transport	0.00	0.00	0.00	0.00	0.80	0.30	0.00	0.10	0.00	0.30
	Efficiency	0.00	0.00	0.00	0.00	0.00	0.80	0.00	0.10	0.20	0.80
	Health	0.00	0.00	0.00	0.00	0.00	0.00	0.80	0.20	0.10	0.10
	Embedded	0.00	0.00	0.00	0.00	0.00	0.00	0.00	0.90	0.10	0.10
	Internet	0.00	0.00	0.00	0.00	0.00	0.00	0.00	0.00	0.70	0.30
	Energy	0.00	0.00	0.00	0.00	0.00	0.00	0.00	0.00	0.00	0.80
OSim@20	Networks	0.85	0.25	0.35	0.40	0.15	0.30	0.20	0.20	0.70	0.30
	Systems	0.00	0.75	0.25	0.20	0.20	0.35	0.25	0.40	0.30	0.35
	Software	0.00	0.00	0.65	0.35	0.20	0.35	0.25	0.30	0.40	0.35
	Services	0.00	0.00	0.00	0.65	0.10	0.30	0.30	0.20	0.40	0.30
	Transport	0.00	0.00	0.00	0.00	0.75	0.40	0.15	0.20	0.20	0.40
	Efficiency	0.00	0.00	0.00	0.00	0.00	0.75	0.25	0.20	0.25	0.75
	Health	0.00	0.00	0.00	0.00	0.00	0.00	0.85	0.30	0.20	0.30
	Embedded	0.00	0.00	0.00	0.00	0.00	0.00	0.00	0.85	0.25	0.25
	Internet	0.00	0.00	0.00	0.00	0.00	0.00	0.00	0.00	0.75	0.35
	Energy	0.00	0.00	0.00	0.00	0.00	0.00	0.00	0.00	0.00	0.75

(continued)

Table 4.6 (continued)

		Networks	Systems	Software	Services	Transport	Efficiency	Health	Embedded	Internet	Energy
OSim@50	Networks	0.74	0.40	0.48	0.46	0.26	0.42	0.24	0.26	0.70	0.42
	Systems	0.00	0.78	0.56	0.56	0.38	0.64	0.48	0.52	0.38	0.64
	Software	0.00	0.00	0.72	0.64	0.32	0.54	0.42	0.40	0.48	0.54
	Services	0.00	0.00	0.00	0.70	0.36	0.52	0.42	0.38	0.50	0.52
	Transport	0.00	0.00	0.00	0.00	0.58	0.54	0.40	0.38	0.34	0.54
	Efficiency	0.00	0.00	0.00	0.00	0.00	0.76	0.46	0.42	0.34	0.76
	Health	0.00	0.00	0.00	0.00	0.00	0.00	0.70	0.38	0.28	0.54
	Embedded	0.00	0.00	0.00	0.00	0.00	0.00	0.00	0.74	0.28	0.54
	Internet	0.00	0.00	0.00	0.00	0.00	0.00	0.00	0.00	0.66	0.52
	Energy	0.00	0.00	0.00	0.00	0.00	0.00	0.00	0.00	0.00	0.76
Kendall's τ	Networks	0.88	0.43	0.61	0.58	0.52	0.51	0.51	0.56	0.77	0.51
	Systems	0.00	0.82	0.52	0.47	0.47	0.48	0.50	0.61	0.43	0.48
	Software	0.00	0.00	0.84	0.71	0.55	0.57	0.55	0.61	0.66	0.57
	Services	0.00	0.00	0.00	0.87	0.59	0.52	0.54	0.55	0.64	0.52
	Transport	0.00	0.00	0.00	0.00	0.83	0.52	0.47	0.53	0.50	0.52
	Efficiency	0.00	0.00	0.00	0.00	0.00	0.85	0.46	0.52	0.49	0.85
	Health	0.00	0.00	0.00	0.00	0.00	0.00	0.87	0.52	0.48	0.44
	Embedded	0.00	0.00	0.00	0.00	0.00	0.00	0.00	0.85	0.53	0.51
	Internet	0.00	0.00	0.00	0.00	0.00	0.00	0.00	0.00	0.89	0.47
	Energy	0.00	0.00	0.00	0.00	0.00	0.00	0.00	0.00	0.00	0.85

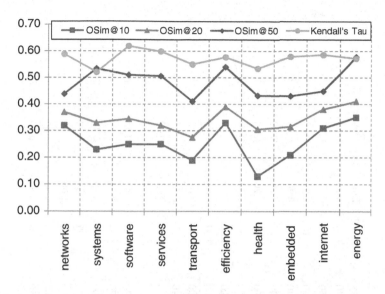

Fig. 4.8 Average values of OSim@10, OSim@20, OSim@50, and Kendall's τ based on Table 4.6

Fig. 4.9 OSim and Kendall's τ for comparison of AHP-based rankings across topics (detailed numbers are available in Table 4.7)

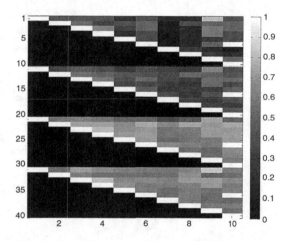

HITS/PageRank-based model that can be fully personalized. The second metric measures organizations' structural importance based on the notion of structural holes. In our approach structural importance is computed with respect to certain topics of interest. Thus, structural importance helps identifying organizations that may be valuable partners for strategic alliances. Combined with authority, this provides a powerful approach for ranking and discovering new partners. Finally, authority and structural importance are systematically combined with cost. For that purpose we utilize AHP to achieve a trade-off among various ranking criteria. The proposed approach delivers very good results and provides more accurate, topic-sensitive results when compared with other ranking techniques.

Table 4.7 OSim and Kendall's τ for comparison of AHP-based rankings across topics

		Networks	Systems	Software	Services	Transport	Efficiency	Health	Embedded	Internet	Energy
OSim@10	Networks	1.00	0.40	0.40	0.40	0.20	0.40	0.20	0.20	0.80	0.40
	Systems	0.00	1.00	0.50	0.30	0.40	0.50	0.10	0.40	0.50	0.50
	Software	0.00	0.00	1.00	0.60	0.30	0.50	0.10	0.20	0.60	0.50
	Services	0.00	0.00	0.00	1.00	0.20	0.40	0.10	0.20	0.60	0.40
	Transport	0.00	0.00	0.00	0.00	1.00	0.40	0.10	0.20	0.30	0.40
	Efficiency	0.00	0.00	0.00	0.00	0.00	1.00	0.10	0.20	0.50	1.00
	Health	0.00	0.00	0.00	0.00	0.00	0.00	1.00	0.30	0.20	0.10
	Embedded	0.00	0.00	0.00	0.00	0.00	0.00	0.00	1.00	0.30	0.20
	Internet	0.00	0.00	0.00	0.00	0.00	0.00	0.00	0.00	1.00	0.50
	Energy	0.00	0.00	0.00	0.00	0.00	0.00	0.00	0.00	0.00	1.00
OSim@20	Networks	1.00	0.35	0.45	0.40	0.30	0.40	0.35	0.30	0.65	0.40
	Systems	0.00	1.00	0.55	0.50	0.45	0.45	0.40	0.55	0.50	0.45
	Software	0.00	0.00	1.00	0.60	0.40	0.55	0.40	0.35	0.55	0.55
	Services	0.00	0.00	0.00	1.00	0.35	0.40	0.35	0.30	0.55	0.40
	Transport	0.00	0.00	0.00	0.00	1.00	0.50	0.30	0.35	0.35	0.50
	Efficiency	0.00	0.00	0.00	0.00	0.00	1.00	0.40	0.30	0.45	1.00
	Health	0.00	0.00	0.00	0.00	0.00	0.00	1.00	0.35	0.30	0.40
	Embedded	0.00	0.00	0.00	0.00	0.00	0.00	0.00	1.00	0.35	0.30
	Internet	0.00	0.00	0.00	0.00	0.00	0.00	0.00	0.00	1.00	0.45
	Energy	0.00	0.00	0.00	0.00	0.00	0.00	0.00	0.00	0.00	1.00

OSim@50	Networks	Systems	Software	Services	Transport	Efficiency	Health	Embedded	Internet	Energy
Networks	1.00	0.58	0.62	0.58	0.50	0.52	0.40	0.46	0.76	0.52
Systems	0.00	1.00	0.74	0.74	0.68	0.78	0.68	0.78	0.72	0.78
Software	0.00	0.00	1.00	0.82	0.70	0.72	0.60	0.64	0.74	0.72
Services	0.00	0.00	0.00	1.00	0.72	0.74	0.60	0.62	0.72	0.74
Transport	0.00	0.00	0.00	0.00	1.00	0.72	0.54	0.60	0.62	0.72
Efficiency	0.00	0.00	0.00	0.00	0.00	1.00	0.68	0.66	0.62	1.00
Health	0.00	0.00	0.00	0.00	0.00	0.00	1.00	0.58	0.54	0.68
Embedded	0.00	0.00	0.00	0.00	0.00	0.00	0.00	1.00	0.56	0.66
Internet	0.00	0.00	0.00	0.00	0.00	0.00	0.00	0.00	1.00	0.62
Energy	0.00	0.00	0.00	0.00	0.00	0.00	0.00	0.00	0.00	1.00

Kendall's τ	Networks	Systems	Software	Services	Transport	Efficiency	Health	Embedded	Internet	Energy
Networks	1.00	0.56	0.72	0.67	0.63	0.62	0.60	0.66	0.86	0.62
Systems	0.00	1.00	0.62	0.54	0.57	0.57	0.57	0.70	0.50	0.57
Software	0.00	0.00	1.00	0.82	0.67	0.68	0.65	0.73	0.74	0.68
Services	0.00	0.00	0.00	1.00	0.70	0.62	0.63	0.65	0.72	0.62
Transport	0.00	0.00	0.00	0.00	1.00	0.64	0.57	0.64	0.58	0.64
Efficiency	0.00	0.00	0.00	0.00	0.00	1.00	0.55	0.63	0.56	1.00
Health	0.00	0.00	0.00	0.00	0.00	0.00	1.00	0.62	0.56	0.55
Embedded	0.00	0.00	0.00	0.00	0.00	0.00	0.00	1.00	0.60	0.63
Internet	0.00	0.00	0.00	0.00	0.00	0.00	0.00	0.00	1.00	0.56
Energy	0.00	0.00	0.00	0.00	0.00	0.00	0.00	0.00	0.00	1.00

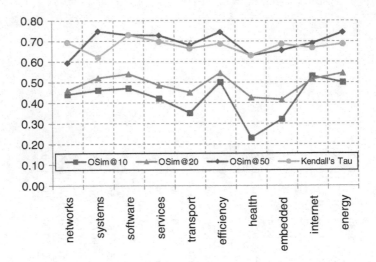

Fig. 4.10 Average values of OSim@10, OSim@20, OSim@50, and Kendall's τ based on Table 4.7

In our future work we will study the application of online formation algorithms [35] to scientific collaboration networks to suggest competitive alliances and consortia. The metrics used in the formation algorithm to rank partners will be based on the techniques as presented in this work.

References

1. D. H. Sonnenwald, B. Cronin, Anonymous, Scientific collaboration: A synthesis of challenges and strategies, in: Annual Review of Information Science and Technology, Vol. 4th, Information Today, 2007, pp. 2–37.
2. F. Fu, C. Hauert, M. A. Nowak, L. Wang, Reputation-based partner choice promotes cooperation in social networks, Phys. Rev. E 78 (2008) 026117. doi:10.1103/PhysRevE.78.026117.
3. C. S. Wagner, L. Leydesdorff, Network structure, self-organization, and the growth of international collaboration in science, Research Policy 34 (10) (2005) 1608–1618. doi:10.1016/j.respol.2005.08.002.
4. Y. Ding, Scientific collaboration and endorsement: Network analysis of coauthorship and citation networks, Journal of Informetrics 5 (1) (2011) 187–203. doi:10.1016/j.joi.2010.10.008.
5. R. Guns, Y. Liu, D. Mahbuba, Q-measures and betweenness centrality in a collaboration network: a case study of the field of informetrics, Scientometrics 87 (1) (2011) 133–147. doi:10.1007/s11192-010-0332-3. URL http://dx.doi.org/10.1007/s11192-010-0332-3
6. M. E. J. Newman, Coauthorship networks and patterns of scientific collaboration, Proceedings of the National Academy of Sciences of the United States of America 101 (Suppl 1) (2004) 5200–5205. arXiv:http://www.pnas.org/content/101/suppl.1/5200.full.pdf+html, doi: 10.1073/pnas.0307545100.

7. N. Lavrac, P. Ljubic, T. Urbancic, G. Papa, M. Jermol, S. Bollhalter, Trust modeling for networked organizations using reputation and collaboration estimates, Systems, Man, and Cybernetics, Part C: Applications and Reviews, IEEE Transactions on 37 (3) (2007) 429–439. doi:10.1109/TSMCC.2006.889531.
8. S. Milojevií, Modes of collaboration in modern science: Beyond power laws and preferential attachment, Journal of the American Society for Information Science and Technology 61 (7) (2010) 1410–1423. doi:10.1002/asi.v61:7.
9. L. Page, S. Brin, R. Motwani, T. Winograd, The pagerank citation ranking: Bringing order to the web, Tech. rep., Stanford University (1998).
10. J. Kleinberg, Authoritative sources in a hyperlinked environment, Journal of the ACM 46 (1999) 668–677.
11. D. Schall, Measuring contextual partner importance in scientific collaboration networks, Journal of Informetrics.
12. L. M. Camarinha-Matos, H. Afsarmanesh, Collaborative networks, in: PROLAMAT, 2006, pp. 26–40.
13. S. Goyala, F. Vega-Redondo, Structural holes in social networks, Journal of Economic Theory 137 (1) (2007) 460–492.
14. W. Tsai, Social capital, strategic relatedness, and the formation of intra-organizational strategic linkages, Strategic Management Journal 21 (9) (2000) 925–939.
15. M. S. Granovetter, The strength of weak ties, The American Journal of Sociology 78 (6) (1973) 1360–1380.
16. R. S. Burt, Structural holes: The social structure of competition., Harvard University Press, 1992.
17. R. S. Burt, Structural holes and good ideas, American Journal of Sociology 110 (2) (2004) 349–399.
18. J. Kleinberg, S. Suri, E. Tardos, T. Wexler, Strategic network formation with structural holes, SIGecom Exchanges 7 (3).
19. L. Backstrom, D. Huttenlocher, J. Kleinberg, X. Lan, Group formation in large social networks: membership, growth, and evolution, in: Proceedings of the 12th ACM SIGKDD international conference on Knowledge discovery and data mining, KDD '06, ACM, New York, NY, USA, 2006, pp. 44–54.
20. T. Haveliwala, S. Kamvar, G. Jeh, An analytical comparison of approaches to personalizing pagerank, Tech. rep., Stanford University (2003).
21. T. H. Haveliwala, Topic-sensitive pagerank, in: Proceedings of the 11th international conference on World Wide Web, WWW '02, ACM, New York, NY, USA, 2002, pp. 517–526. doi:10.1145/511446.511513. URL http://doi.acm.org/10.1145/511446.511513
22. G. Jeh, J. Widom, Scaling personalized web search, in: Proceedings of the 12th international conference on World Wide Web, WWW '03, ACM, New York, NY, USA, 2003, pp. 271–279. doi:10.1145/775152.775191. URL http://doi.acm.org/10.1145/775152.775191
23. P. Berkhin, Survey: A survey on pagerank computing., Internet Mathematics 2 (1) (2005) 73–120.
24. S. Chakrabarti, Dynamic personalized pagerank in entity-relation graphs, in: Proceedings of the 16th international conference on World Wide Web, WWW '07, ACM, New York, NY, USA, 2007, pp. 571–580. doi:10.1145/1242572.1242650. URL http://doi.acm.org/10.1145/1242572.1242650
25. D. Fogaras, K. Csalogany, B. Racz, T. Sarlos, Towards scaling fully personalized pagerank: Algorithms, lower bounds, and experiments, Internet Mathematics 2 (3) (2005) 333–358.
26. H. Deng, M. R. Lyu, I. King, A generalized co-hits algorithm and its application to bipartite graphs, in: Proceedings of the 15th ACM SIGKDD international conference on Knowledge discovery and data mining, KDD '09, ACM, New York, NY, USA, 2009, pp. 239–248. doi:10.1145/1557019.1557051. URL http://doi.acm.org/10.1145/1557019.1557051
27. K. Berberich, M. Vazirgiannis, G. Weikum, T-rank: Time-aware authority ranking, in: S. Leonardi (Ed.), Algorithms and Models for the Web-Graph, Vol. 3243 of Lecture Notes in Computer Science, Springer Berlin Heidelberg, 2004, pp. 131–142.

28. D. Schall, Expertise ranking using activity and contextual link measures, Data Knowl. Eng. 71 (1) (2012) 92–113.
29. D. Schall, Service Oriented Crowdsourcing: Architecture, Protocols and Algorithms, Springer Briefs in Computer Science, Springer New York, New York, NY, USA, 2012.
30. T. L. Saaty, Decision making with the analytic hierarchy process, International Journal of Services Sciences 1 (2008) 83–98.
31. F. Munisteri, ICT statistical report for annual monitoring 2011, http://ec.europa.eu/digital-agenda/sites/digital-agenda/files/stream_2012_0.pdf (Feb. 2012).
32. A. Y. Ng, A. X. Zheng, M. I. Jordan, Stable algorithms for link analysis, in: Proceedings of the 24th annual international ACM SIGIR conference on Research and development in information retrieval, SIGIR '01, ACM, New York, NY, USA, 2001, pp. 258–266.
33. P. Ferrari, A method for choosing from among alternative transportation projects, European Journal of Operational Research 150 (1) (2003) 194–203.
34. A. Certa, M. Enea, T. Lupo, Electre iii to dynamically support the decision maker about the periodic replacements configurations for a multi-component system, Decis. Support Syst. 55 (1) (2013) 126–134.
35. A. Anagnostopoulos, L. Becchetti, C. Castillo, A. Gionis, S. Leonardi, Online team formation in social networks, in: Proceedings of the 21st international conference on World Wide Web, WWW '12, ACM, New York, NY, USA, 2012, pp. 839–848. doi:10.1145/2187836.2187950.

Chapter 5
Social Broker Recommendation

Abstract In this chapter we propose novel socially-based models for the composition of Professional Virtual Communities (PVC). We focus on the notion of brokers who act as intermediaries between separated communities. We introduce a broker discovery and ranking approach utilizing a link-based broker importance model. We evaluate our approach through a service-oriented testbed and real community data obtained from the European Union's FP7 research program.

5.1 Virtual Organizations

The rapid advancement of ICT-enabled infrastructure has fundamentally changed how businesses and companies operate. Global markets and the requirement for rapid innovation demand for alliances between individual companies [9]. Virtual Organizations (VO) are an important concept in such dynamic environments. Based on [18], a virtual organization can be defined as follows: *an inter-organizational virtual organization is a temporary network organization, consisting of independent enterprises (organizations, companies, institutions, or specialized individuals) that come together swiftly to exploit an apparent market opportunity. The enterprises utilize their core competencies in an attempt to create a best-of-everything organization in a value-adding partnership, facilitated by Information and Communication Technology (ICT). As such, virtual organizations act in all appearances as a single organizational unit.*

Web services and service-oriented computing offer well established standards and techniques to model and implement interactions spanning multiple organizations. Collaborative service-based systems are typically knowledge intensive covering complex interactions between people and software services. In such ecosystems, flexible interactions commonly take place in different organizational units. The challenge is that top-down composition models are difficult to apply in constantly changing and evolving service-oriented collaboration system. There are two major obstacles hampering the establishment of seamless communications and collaborations across organizational boundaries:

- the dynamic discovery and composition of resources, people and services, and
- flexible interactions between people located in different departments or companies.

© Springer International Publishing Switzerland 2015
D. Schall, *Social Network-Based Recommender Systems*,
DOI 10.1007/978-3-319-22735-1_5

Principles found in social network theory are promising candidate techniques to assist in the formation process and to support flexible and evolving interaction patterns in cross-organizational environments. In social networks, relations and interactions typically emerge freely and independently without restricted paths and boundaries. Research in social sciences has shown that the resulting social network structures allow for relatively short paths of information propagation (the small-world phenomenon [46]). While this is true for autonomously forming social networks, the boundaries of collaborative networks are typically restricted due to organizational units and fragmented areas of expertise.

We propose social network principles to bridge segregated collaborative networks. The theory of structural holes is based on the idea that individuals can benefit from serving as intermediaries between others who are not directly connected [7, 8]. Thus, such intermediaries can potentially broker information and aggregate ideas arising in different parts of a network [21, 38, 40]. The novelty of the present work is the combination of social principles for the discovery of brokers and service-oriented collaboration infrastructure through Human-Provides Services (HPS) [34, 39]. HPS supports a flexible interaction model suitable for cross-organizational collaboration.

In this chapter, the following key contributions are presented:

- We introduce concepts and techniques for the discovery of brokers in virtual communities. Here we present *SBQL*, a query language for discovering and ranking brokers. SBQL evolved from our initial definition of social network query language (see [40]). SBQL provides a better integration into our mining and ranking framework and allows for fuzzy matches with ranked results, which was not possible in our previous query language.
- In this work we present a novel link-based broker importance model. The importance model is based on the idea of hubs and authorities in Web-based environments as introduced by [20]. Our proposed model can be personalized to account for topic-sensitive rankings.
- We performed various experiments and tests including performance tests of our BrokerQL implementation using an integrated Web services test environment. In addition, we obtained data from the EU FP7 ICT research program (see [27]) to test the broker ranking approach in a real virtual collaboration environment.

To address the challenges related to broker discovery and compositions in PVCs, we apply the following overall methodology:

- We model a PVC as a community consisting of experts who interact and collaborate by the means of ICT to perform work. Service-oriented technologies (Web services, RESTful services, etc.) provide the technical infrastructure to perform collaborations.

- The emergence and evolution of social trust is modeled by considering various metrics. In contrast to a common security perspective on trust, we define social trust to rely on the interpretation of previous collaboration behavior. Considering social trust is essential to effectively guide interactions.
- We define patterns for broker discovery in virtual communities. The notion of brokers is derived from the well-established theory of structural holes. We define the persistent exogenous interaction pattern and the triadic exogenous interaction pattern.
- To actually discover brokers in VOs and PVCs, we define the BrokerQL language and its syntax. A set of BrokerQL applications and examples are discussed to illustrate the features of BrokerQL.
- BrokerQL includes a novel broker importance ranking algorithm that is mathematically modeled and discussed in depth. Thus, not only matching of relevant brokers is performed, but also ranking based on social/collaborative network information.
- All presented concepts are implemented in software and tested through rich experiments. In addition, the quality of broker rankings is validated by using a real dataset reflecting collaborations in virtual environments.

This chapter is structured as follows. Discussion regarding existing literature is presented in Sect. 5.2. In Sect. 5.3, we introduce supporting concepts to realize flexible interactions and the selection of brokers. In Sect. 5.4, we present a motivating scenario for the discovery of brokers. We introduce the social broker query language in Sect. 5.5 followed by the definition of the broker ranking model in Sect. 5.6. Our evaluation results are discussed in Sect. 5.7. Finally, the chapter is concluded in Sect. 5.8.

5.2   Background in Distributed Organizations

Here we review background literature and cluster discussions into relevant topics. We start with discussions related to virtual organizations and communities.

- **Virtual Organizations** (VO) can be studied from various angles. From a management point of view, the concept of virtual organizations that are supported by ICT has been widely studied (for example, see [9]). The central goal if this work is to study VOs and the formation thereof from a *social network* point of view. One of the most interesting ideas in the social sciences is the notion that entities are embedded in webs of social relations and interactions [6]. The same holds true for today's VO landscape where a globally distributed organizations form temporary alliances to work on joint projects. Collaborative non-hierarchical business networks enable, for example, the execution of complex product manufacturing processes [42]. Formation of globally distributed teams or VOs is an important and challenging problem [4]. It has been reported that, for example, team assembly mechanisms determine both the structure of the collaboration network and performance for teams [14]. In addition, social model for socio-technical performance have been presented [47]. Thus, it is important to select a formation strategy carefully according to constraints and objectives [29]. An interesting application of socially-based formation principles is social product development in VOs [5].

- **Structural Holes.** Here we adopt the theory of structural holes as developed by [7]. The theory is based on the idea that individuals can benefit from serving as intermediaries between others who are not directly connected [8]. Such intermediaries can potentially broker information and aggregate ideas arising in different parts of a network [21]. A number of studies have shown that structural holes positively relate to a range of social success indicators [1, 7, 8, 31]. Lou and Tang [23] define the problem of mining top structural hole spanners in large-scale social networks.

- **Social Trust.** In addition to the notion of brokers in collaborative networks, we build upon research in the area of *social trust*. A wide range of computational trust models have been proposed by, for example, [2, 19, 26]. Here we focus on social trust [11, 44, 48] that relies on user interests and collaboration behavior. A social network of people connected by trust relations is a fundamental building block in many of today's most successful e-commerce and recommendation systems [13].

- **Human Provided Services.** Based on social principles such as brokers and social trust, we provide technical concepts to support the discovery and ranking of brokers. We go beyond social concepts only and propose the (semi-)automatic formation of broker-based communities and service-oriented interactions. Service-oriented concepts help to rapidly setup and operate VOs. Here the concept of *Human-Provided Services* (HPS) [34, 39] is adopted which supports flexible service-oriented collaboration across multiple organizations and domains. HPS not only enables flexible interactions but also provides the base infrastructure for interaction mining and link-based ranking techniques [35, 36]. In contrast to our previous work, we propose a domain specific query language that is geared towards the requirements of broker discovery and ranking. Here we introduce *BrokerQL*, which is a query language specifically suited for broker discovery in socially-based collaboration networks. BrokerQL is based on an SQL-style syntax targeted at social and collaborative networks.

- **Querying Social Network Data.** To date, to the best of our knowledge there is no generic solution to query, rank, and select brokers for assembling VOs or groups in social networks. Some proposals for graph databases (e.g., GraphDB [15]) have features to deal with social network data [41], but lack advanced features for broker discovery. SPARQL [45] is a generic language to query semantic data such as RDF graphs, but lacks some important capabilities. SPARQL (1) does not support negation ("A does not know B"), which is fundamental in VO formation (competition or conflicting interests), (2) expressing network properties such as path-length is not straightforward, (3) predicates cannot have properties (e.g. "A trusts B with trust level high") and finally (4) fuzzy matches with ranked result is difficult. A query language for social networks has been proposed in [32].

The language in [32] has some similarities with SBQL (e.g., path functions), however, without supporting the discovery of complex sub communities based on metrics and interaction mining techniques.

5.3 Hybrid Compute Environment

We adopt various concepts to realize the aforementioned flexible collaboration communities, and consider various mechanisms to enable brokering of requests, including flexible collaboration models, automatic management of social trust relations and modeling of broker patterns and behavior.

5.3.1 Human-Provided Services

An activity model attempts to structure loosely coupled collaborations in service-oriented systems (see [34] for details). Examples of collaborative activities at various levels of granularity are "sending emails", "reviewing a paper", "organizing a workshop", and "managing a multi-national research project". A single activity provides the basic collaborative data. It describes the work to be done and lists the involved people. This data is often sufficient in static collaborative settings where all members are aware of the overall working environment. In dynamic and distributed collaborative environments, one has to explicitly model the embedding of a single activity in the overall collaboration context. The context contains the structure of activities, dependencies between activities, the temporal flow of activities, and history of activity changes. This provides the core structure of collaborative work. In addition, the collaboration context describes the involvement of members, their roles, required and applied skills, work artifacts, and resources. An expressive activity model needs to support such relations. Existing activity-based tools provide limited means to connect activities, members, and resources. Web services play a fundamental role in supporting flexible, cross-organizational collaboration scenarios.

To support human interactions in a service-oriented manner, we have designed and implemented the HPS framework [34]. HPS enhances the traditional "SOA-triangle" approach by enabling people to provide services using the very same technology as implementations of software-based services (SBS) use. In contrast to technologies such as BPEL4People [3], HPS treats people in SOA as first-class citizens letting people define their own services and provide these services to the community. HPS closely resembles a *crowdsourcing* approach [36] where the collective power of people is used to solve problems that services implemented in software cannot yet solve.

The three essential steps performed when using the HPS framework are illustrated in the following. The conceptual model is also depicted by Fig. 5.1.

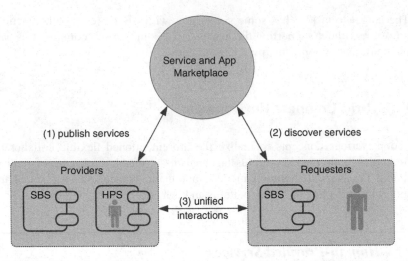

Fig. 5.1 HPS discovery and interaction model overview: (1) publish HPS or SBS to service registry, (2) search and discover HPS and/or SBS, and (3) interact with service provider

1. *Publish services.* The user can create an HPS and publish the service in a *Service and App Marketplace*. Publishing a service is as simple as posting a blog entry on the Web. It is the association of the user's profile with an activity depicted as a service.
2. *Discover services.* The requester can perform a search query to find Human-Provided Services or Apps that are based on Human-Provided Services. Recommendations are given to find the most relevant HPS based on, for example, the expertise of the user providing the service [35].
3. *Unified interactions.* The framework supports both "machine interactions" between SBS and HPS (a human computation scenario) and human interactions between a human requester and an HPS. Apps offer the interfaces to support such kind of interactions.

5.3.2   Emergence and Evolution of Social Trust

In contrast to a common security perspective on trust, we define social trust to rely on the interpretation of previous collaboration behavior and additionally consider the similarity of dynamically adapting interests [11, 44]. Especially in collaborative environments, where users are exposed to higher risks than in common social network scenarios [10], and where business is at stake, considering social trust is essential to effectively guide interactions [24].

In this work, we define social trust as follows [11, 12, 26, 44]:

Trust reflects the expectation one actor has about another's future behavior to perform given activities dependably, securely, and reliably based on experiences collected from previous interactions.

Not only service interactions, but also human interactions may rely on standard protocols such as SOAP (e.g., see HPS [34] and BPEL4People [3]). These protocols are well supported by a wide variety of software frameworks. This fact enables the adoption of various available monitoring and logging tools to observe interactions in service-oriented systems. Various metrics can be calculated by analyzing interaction logs. Relation metrics describe the links between actors by accounting for (1) recent interaction behavior, (2) profile similarities (e.g., interest or skill similarities), (3) social and/or hierarchical structures (e.g., role models). However, we argue that social trust relations largely depend on personal interactions. We model a community of actors with their social relations as a directed graph, where the nodes denote network members, and edges reflect (social) relations between them. Since interaction behavior is usually not symmetric, actor relations are represented by *directed links*.

The fundamental approach to automatic interaction-based trust inference is depicted in Fig. 5.2. As motivated in the introduced use case, people interact to perform their tasks. This work is modeled as activities, that describe the type and goal of work, temporal constraints, and used resources. As interactions take place in the context of specific activities (Fig. 5.2a), they can be categorized and weighted. Interaction logs are used to infer metrics that describe the relation between individual actors (Fig. 5.2b), such as their behavior in terms of availability and reciprocity.

Our approach considers the diversity of trust by enabling the flexible aggregation of various interaction metrics that are determined by observing ongoing collaborations. Finally, available relation metrics are weighted, interpreted, and composed by a rule engine. The result describes trust between the actors with respect to scopes (Fig. 5.2c). For instance, trust relations in a scope "scientific dissemination"

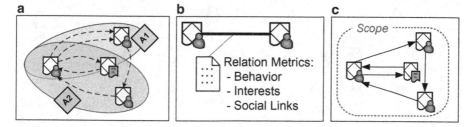

Fig. 5.2 Trust emerging from interactions: (**a**) interaction patterns shape the behavior of actors in context of activities; (**b**) rewarding of behavior and calculation of interaction metrics; (**c**) inference in scopes by interpretation of metrics

could be interpreted from interaction behavior of actors in a set of "paper writing" activities. For detailed mechanisms on trust modeling and inference see [44].

5.4 Expert Communities

Here we discuss a PVC environment to introduce our concepts and to demonstrate the role of social relations and the emergence of social trust. Also, we motive the need for brokers in PVC environments. A PVC is a virtual community consisting of experts who interact and collaborate by the means of information and communication technologies to perform work [9]. In today's collaboration environments, service-oriented technologies (Web services, RESTful services, etc.) are increasingly used to realize a collaboration infrastructure suitable for PVCs. An important aspect is the availability of a wide variety of tools and frameworks to implement service-oriented systems.

5.4.1 Collaboration Scenario

The support of loose coupling, advanced discovery, dynamic binding and various composition mechanisms makes SOA the ideal grounding for Web-enabled PVCs. Let us discuss an actual collaboration scenario in PVCs as depicted in Fig. 5.3.

Various member groups collaborate in the context of five different activities a_1, a_2, a_3, a_4 and a_5 (see Fig. 5.3a). These groups intersect because group members may participate in different activities at the same time. The color of the activity determines the context and expertise areas an activity is related to. Such activities are, for example, the creation of new design specifications or the discussion of emerging technology standards. Activities are a concept to structure information

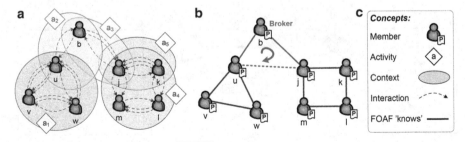

Fig. 5.3 Collaboration model for service-oriented PVCs: (**a**) interactions between PVC members are performed in the context of activities; (**b**) social relations and profile areas emerge based on interactions; (**c**) concepts in the model

in flexible collaboration environments [25], including the goal of the ongoing tasks, involved actors, and utilized resources such as documents or services [34]. Activities are either assigned from the outside of a community, e.g., belonging to a higher-level business process, or emerge by identifying collaboration opportunities.

To achieve the goals of activities, the PVC members use SOA technology to interact in the context of the currently performed activities. In the depicted scenario, we the concept of Human-Provided Services (HPS) [39] and the HPS framework [34] is used is used to allow human participation in a service-oriented manner. That is, humans can provide their capabilities and skills as *services* to enable human interactions through a standardized message exchange format (i.e., SOAP). Instead of implementing services in software, services are provided by human actors. The novelty of the approach is that HPS enables a seamless service-oriented infrastructure consisting of human and software services. At the technical level, all messages including SOAP-based messages are logged for later analysis of communities and broker discovery.

Relations emerge from interactions as illustrated in Fig. 5.3b and are bound to particular scopes. We model the interaction context with tags and keywords and compose similar activities to build trust scopes. To infer trust, we aggregate interactions that occurred in a pre-defined scope, calculate metrics (numeric values describing prior interaction behavior), and interpret them in terms of reliability, dependability and success. In the given scenario, a scope comprises trust relations between PVC members regarding help and support in different expertise areas (reflected by tags of exchanged messages). Through analyzing the interaction context (i.e., using message tags), we determine a user's centers of interests. Frequently used keywords are stored in the actors' profiles (see symbol P) and later used to determine their interests and expertise areas.

5.4.2 Brokerage and Composition

Consider a scenario in the given PVC in Fig. 5.3b where u wants to set up an activity that requires at least one additional expert from the *brown* $\{u, v, w\}$ and *blue* domain $\{j, k, l, m\}$. Since u personally knows v and w from previous collaborations (reflected by social relations), u is well-connected to the *brown* expertise area; but u does not know any member of the *blue* domain. However, in the scenario u collaborated with b in the *green* domain, who is connected to j. Hence, b could act as a broker and forward requests or invitations to join u's current activity to j. We argue that establishing personal contacts in socially-based environments is of high importance compared to the traditional SOA domain, where services are mostly composed based on their properties (i.e., features and QoS) only.

Interaction mining techniques support the discovery of emerging social relations. These relations have major impact on future collaborations such as:

- *Supporting the Formation of Expert Groups.* Successful previous formations of actors should by "recorded" to actively facilitate future collaborations. Thus, tight trust relations based on interactions can be converted to social relations.
- *Controlling Interactions and Delegations.* Discovery and interactions between members can be based on social relations. People tend to favor help and support requests from well-known members instead of receiving requests from any third (personally unknown) parties.
- *Establishment of new Social Relations.* The emergence of new personal relations is actively facilitated through brokers. The introduction of new partners through brokers (e.g., *b* introduces *u* and *j* to each other) leads to future trustworthy compositions.

5.4.3 Broker Patterns and Policies

Brokers as outlined in the beginning of Sect. 5.4 differ from the other actors in the environment by their mediation capabilities. A broker acts as an intermediary between two previously separated collaboration teams. It is thus essential that it gathers frequently demanded contacts and maintains its relations. However, if demand decreases the broker must find and establish new relations.

Broker Patterns The remainder of this section describes the broker patterns to establish connections between requesters and third-party services. Brokers can typically have different behavior when delegating requests. In particular, as shown in Fig. 5.4, we distinguish:

- *Persistent Exogenous Interaction Pattern* (Fig. 5.4a) where any requests and responses are forwarded by the broker and the actually interacting nodes are shielded from each other.
- *Triadic Exogenous Interaction Pattern* (Fig. 5.4b) where the broker encourages receivers of messages to establish direct connections to the initially requesting nodes, and therefore, actively facilitates the emergence of trust relations.

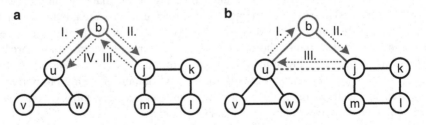

Fig. 5.4 Exogenous broker behavior patterns. (**a**) Persistent interaction pattern. (**b**) Triadic interaction pattern

We argue that both types of interaction patterns are applied in today's social and collaborative environments. The broker may favor one over the other pattern due to various reasons. On the one side, controlling the flow of interactions between personally unknown actors can strengthen a broker's reputation. On the other side, establishing direct relations can significantly reduce a broker's working load.

Behavior Policies Policies along with the queries provide general or explicit rule-sets to define expectations on path requests and responses. While the queries in the presented scenario are limited to more static lookup properties such as, competences and knowledge fields a policy-driven management of the interactions also allows to consider the relation between a query and the current context. A request policy limits the request and result by an according rule-set. Rules include a condition part possibly matching both current query and context and a decision part that filters the results. Thus, a request triggered policy could state that a requester wishes, e.g., a maximum/minimum path length to the knowledge source, a minimum trust relation from her/his side to the broker or also from the broker to the expert, hard completion deadlines for the subsequently delegated work, etc. Thereby, rules or entire policies can by mandatory or depending on the importance and context desirable for the issued query. Response policies at the broker comprise filter rules for the original query's response paths. Depending on the situation the broker might want to adapt the response. The history of past interactions with the requester, contracts, or independent from those, current changes and events, e.g., interest shifts at or availability of the requested experts require dynamic adaptations of the response paths. These dynamics might also require that the policies' rules are updated accordingly. In order to solve the challenge of updating these situation dependent policies means of querying the network structure are necessary.

Our proposed Social Broker Query Language (SBQL) helps to solve the issues by providing a syntax that can find paths to resources connected to query and policy constrains. We introduce the technical details of SBQl in the following section.

5.5   SBQL Syntax

In this section we define the key elements of the SBQL. The language is inspired by an SQL-like syntax. It is important to note that SBQL operates on a graph structure composed of a set of nodes and edges.

- Select: A Select statement retrieves nodes and edges in graph G as well as aggregates of graph properties (for example, properties of a set of nodes).
- From: While traditional relational databases operate on tables, SBQL uses the From clause to perform queries on a graph G.
- Where: A Where clause specifies filters and policies upon nodes, edges, and paths.

Table 5.1 Important SBQL language elements

Element	Description
satisfy	Requires that a given condition is fulfilled by a set of nodes or edges
as	Creates an alias for groupings of nodes, edges, or paths
<all>	Retain all nodes/edges/subgraphs satisfying a given condition
[ ]	An expression to satisfy conditions for exactly one [1], one to m [1..m], or one to many [1..*] nodes or edges

```
1  Input: Graph G, var source = {n₁,n₂,...,nᵢ},
2        var target = {nⱼ,nⱼ₊₁,...,nⱼ₊ₘ}
3  Output: List of brokers
4
5  Select node From (
6    ( Select distinct(node) From G
7    Where
8        /* At least one in source 'knows' node */
9        ( [1..*] n in source ) satisfy
10       Path (n to node) as P1 With P1.length = 1 ) as G1,
11   ( target ) as G2
12 )
13 Where
14   /* Retain all nodes that satisfy path filter */
15   ( <all> n in G1.nodes ) satisfy
16     /* Path to any in G2.nodes */
17     Path (n to [1..*] G2.nodes) as P2 With P2.length = 1
18   and
19     /* Retain all edges that satisfy edge filter */
20   ( <all> e in G1.edges ) satisfy
21     (e.relation = EPredicates.BIDIRECTIONAL) and
22     (e.trust >= MTrust.MEDIUM)
23
24 Order by node
```

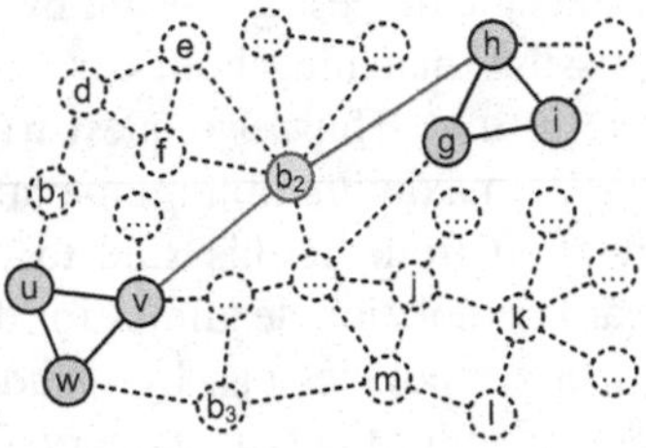

Fig. 5.5 SBQL query: find broker to connect two predefined communities

Note that SBQL is not a general purpose graph (social network) query language but rather a domain specific language to discover and rank brokers in PVCs as well as social and collaborative networks. Table 5.1 lists important SBQL language elements to query and filter graphs.

To give intuitive examples, we present a set of SBQL queries along with their meaning considering a graph G and a set of subgraphs $G' \subseteq G$. We structure discussions related to a query into four essential steps:

- **R** the basic requirements/goal of a query
- **A** the approach that is taken
- **O** the output of the query
- **D** the detailed description of the query

5.5.1 Connecting Predefined Communities

Consider two initially disconnected communities (sets of nodes in Fig. 5.5) depicted as variables var source = $\{n_1, n_2, \ldots, n_i\}$ and var target = $\{n_j, n_{j+1}, \ldots, n_{j+m}\}$ residing in the graph G.

R1: The goal is to find a broker connecting disjoint sets of nodes (i.e., not having any direct links between each other).

A1: Two subgraphs G1 and G2 are created to determine brokers which connect the source community $\{u,v,w\}$ with the target community $\{g,h,i\}$ (i.e., see From construct).

O1: The output of the query is (the example shown in Fig. 5.5) a list of brokers connecting $\{u,v,w\}$ and $\{g,h,i\}$. The lines 1–3 specify the input/output parameters of the query.

D1: As a first step, a (sub)select is performed using the statement as shown by the lines 6–10. The statement distinct (node) means that a set of unique brokers shall be selected based on the condition denoted as the Where clause with a filter (lines 9–10).

The term "[1..*] n in source", where source is the set of nodes passed to the query as input argument, means that at least one node $n \in G$ must satisfy the subsequent condition. Here the condition is that the node n has a link (i.e., through knows relations) to the source set of nodes. This is accomplished by using the Path function that checks whether a link between two nodes exists (the argument "(n to node)"). The path alias is used to specify additional constraints such as the maximum path length between nodes (here "P1 With P1.length = 1"). The second step is to create an alias G2 for the target community $\{g,h,i\}$.

By using the aliases G1 (line 10) and G2 (line 11) further filtering can be performed using the Where clause in line 13. The same syntax is used as previously in the sub-select statement (lines 9–10). The construct <all> retains nodes "n in G1.nodes" (G1 holding the set of candidate brokers) that are connected to at least one node in the target community G2 with direct links ("P2 with P2.length = 1"). Further filtering is performed by defining lines 18–20. Here, the target community $\{g,h,i\}$ must have edges between each other that are bidirectional. In our graph representation, this means that each relation has to be interpreted as, for example, h knows g and g knows h. A set of different metrics is established in our system. A specific type of metric (e.g., trust) is denoted by the namespace MTrust. In the specified query, each actor in the target community must share a minimum level of trust depicted as "e.trust >= MTrust.MEDIUM". Trust metrics are associated to edges between actors. The term MTrust.MEDIUM is established based on mining data to obtain linguistic representations by mapping discrete values (metrics) into meaningful intervals of trust levels.

The last statement "Order by node" in Fig. 5.5 implies a ranking procedure of brokers. This can be accomplished by using eigenvector methods in social networks such as the PageRank algorithm [30] to establish authority scores (the importance or social standing of a node in the network). The detailed mechanisms of this procedure will be discussed in the following section.

```
1   Input: Graph G, var search = {t₁,t₂,...,tₙ}
2   Output: List of communities
3
4   Select load, nodes from (
5     ( Select distinct(nodes) as G' from G
6     Where
7       ( <all> n in G'.nodes ) satisfy
8         Path (n to [1..*] G'.nodes) as P1
9         With (
10          P1.length = 1 and P1.trust = MTrust.HIGH and
11          ( [1..*] tag in P1.tags ) satisfy
12            (search contains tag)
13        )
14    ) as SG1
15    Where
16      ( <all> G' in SG1 ) satisfy
17        (G'.load <= GMLoad.MEDIUM)
18
19  Order by load asc
```

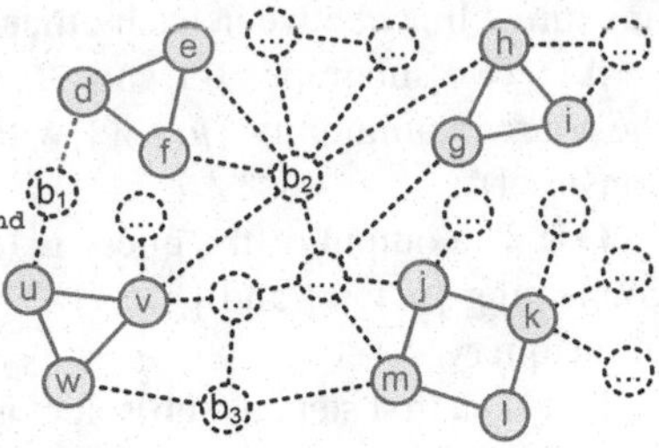

Fig. 5.6 SBQL query: find ranked communities based on search criteria and load metrics

5.5.2   Finding Communities

The broker discovery example in the previous section (depicted by Fig. 5.5) is straightforward because the target community is already specified and passed to the query as var target = $\{n_j, n_{j+1}, \ldots, n_{j+m}\}$. However, in most cases (as highlighted in the introduction example) the target community may not be known beforehand. The next example query eliminates this assumption by showing an approach to find suitable communities based on search criteria (e.g., activity or skill tags).

R2: The goal of the query as specified in Fig. 5.6 is to find sub-communities (or subgraphs) in G that match search criteria.

A2: Search is performed by using a set of distinct tags specified as input parameter var search = $\{t_1, t_2, \ldots, t_n\}$.

O2: The output of the query is a list of communities.

D2: The first step is to perform a (sub)select of distinct communities (see distinct(nodes) as G' in line 5) to obtain non-overlapping groups of community members specified by the lines 5–14. For example, Fig. 5.6 shows four groups of nodes $[\{d,e,f\},\{g,h,i\},\{l,m,j,k\},\{u,v,w\}]$ each of them satisfying the constraints specified in the query. Each node in a specific community must be linked to at least one community member ("Path (n to [1..*] G'.nodes) as P1"). Furthermore, at least one path between nodes with "length = 1" satisfying trust requirements (MTrust.HIGH) must exist in order to consider a node as a community member. Finally, a path must contain the tags specified by the search query (lines 11–12) to ensure that a member has interacted (collaborated) with other members in the context of certain activities.

The alias SG1 provides access to each community. The Where clause applies filtering of communities based on load conditions measured by graph metrics (GMLoad). For example, load conditions G'.load are measured by the number of inbound requests and the number of pending tasks within the community.

```
 1  Input: Graph G, var source = {n₁,n₂,...,nⱼ},
 2         var search = {t₁,t₂,...,tₙ}
 3  Output: List of brokers and communities
 4
 5  Select node, nodes from (
 6      /* Select brokers */
 7      ( /* ... */ ) as G1,
 8      /* Select communities */
 9      ( /* ... */ ) as SG1
10  )
11  Where
12      ( <all> n in G1.nodes ) satisfy
13      /* To one in SG1 */
14      Path (n to [1] SG1) as P1 With P1.length = 1
15
16  Order by node
```

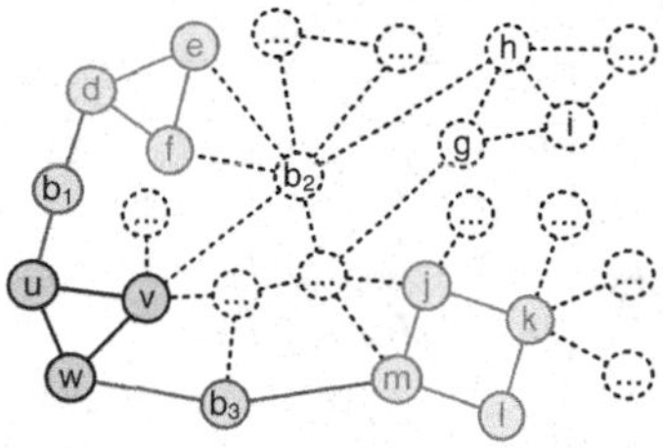

Fig. 5.7   SBQL query: find exclusive brokers to connect two communities

5.5.3   Finding Exclusive Brokers

The final SBQL example is depicted by Fig. 5.7 to combine previously introduced concepts for broker discovery.

R3: The basic idea of this example is to find brokers that are connected to exactly one candidate (target) community.

A3: Communities are retrieved along with brokers. Filtering is applied based on paths to obtain exclusive brokers.

O3: The output of the query are brokers along with communities they are connected to (e.g., b_1, $\{d,e,f\}$).

D3: First, a set of candidate brokers is retrieved and made available via the alias G1 (line 6). This is the same procedure as introduced before (see Fig. 5.5). Second, communities are retrieved and stored in SG1 (line 8). Again, this is based on the same principle as introduced previously in Fig. 5.6. We call brokers connecting exactly one community *exclusive brokers*. This is accomplished by the statements in 11–13 demanding for "n to [1] SG1". The broker b_2 is a non-exclusive broker because it connects multiple communities $\{d,e,f\}$ and $\{g,h,i\}$, thereby making $\{g,h,i\}$ unreachable from the $\{u,v,w\}$ community perspective.

5.6   Broker Ranking

5.6.1   Community Profiles

In contrast to common top-down approaches that apply taxonomies and ontologies to define certain interest and expertise areas, we follow an interest mining approach that addresses the inherent dynamics of flexible collaboration environments. Skills and expertise as well as interests change over time, but are rarely updated if they are managed manually in a registry. Hence, we determine and update them automatically through mining. As discussed before, interactions, e.g., delegation

a   **b**

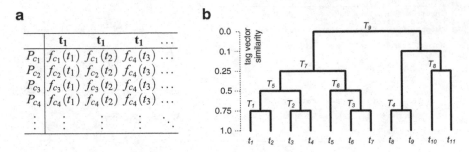

Fig. 5.8 The concept of hierarchical tag clustering: (**a**) tag matrix to determine the co-occurrence (and thus potential similarity) of tags; (**b**) clustering of tag vectors with varying similarity thresholds creates a tag tree

of tasks or requests, are tagged with keywords. As delegation receivers process tasks, our system is able to learn how well people cope with certain tagged tasks; and therefore, able to determine their centers of interests. We calculate *community profiles* by aggregating individual (tag-based) interest profiles.

The community profile P_c in Eq. (5.1) describes the frequencies f_c of the tags $T = \{t_1, t_2, t_3 \dots\}$ that are applied in collaborations in a community c.

$$P_c = \langle f_c(t_1), f_c(t_2), f_c(t_3) \dots \rangle \tag{5.1}$$

Combining multiple community profiles leads to a tag matrix as shown in Fig. 5.8. Tag vectors $\mathbf{t_x}$ describe the usage of a certain tag t_x from a global perspective, i.e., spanning all communities c_i. A common assumption is that co-occurrence of tags reflect their similarity and closeness respectively [22, 43]. We cluster tags based on their similarity. Clustering tags has two major advantages:

- When someone is searching for a particular tag, all similar tags (e.g., synonyms in the same cluster) can be considered in the search process as well to increase the number of results.
- Since our ranking approach uses personalization techniques when searching for brokers, we can significantly reduce the time effort by pre-calculating topic-based broker importance (see later for details).

By comparing the similarity of tag vectors, e.g., using cosine similarity, with varying thresholds, a tree structure is created as depicted in Fig. 5.8. That tree reflects the closeness of single tags and created clusters. For instance tags t_1 and t_2 are merged in cluster T_1. Details are described in [44]. The benefit of this approach in the context of SBQL is that fuzzy matches of communities and/or brokers are possible and topic-based broker importance scores can be calculated at various topic levels (at different levels in the hierarchical topic/tag tree).

5.6.2 Trust Weights

Interactions such as delegations are aggregated to metrics that are interpreted by rules to infer trust. Trust scores are associated to edges (i.e., e.trust) and mapped to trust intervals (e.g., MTrust.MEDIUM). To calculate metrics, the edge weight w_{uv} can be interpreted as how much u trusts v in processing tasks or help and support requests in a reliable manner [Eq. (5.2)]. Specifically, experts' behavior in terms of reliability and task processing successes, are periodically updated with recent captured interaction data.

$$w_{uv} \equiv \frac{\text{succ. delegations from } u \text{ to } v}{\sum_{z \in N(u)} \text{succ. delegations from } u \text{ to } z} \tag{5.2}$$

Reliability and processing success (and thus *social trust*) of tasks/delegations are based on a *task rewarding schema*. Let us assume a human task *ht*. The task has states such as *accepted, inprogress, finished* or *aborted*. Rewards are automatically associated with *ht* to measure the **degree of success**. For example, fast and reliable processing of tasks yields higher rewards, thereby resulting in higher trust in a collaboration partner.

To model task rewards based on temporal task properties (processing time), we use a mathematical function belonging to the family of sigmoid functions with the general form $f(t) = \frac{1}{1+e^{-t}}$ (see Fig. 5.9). Sigmoid functions are typically used to model systems that saturate at large values of t, for example, the processing time of tasks. Let us define the essential properties of the model in the following. The task rewarding function RW based on the task processing time $PT(ht)$ for a given task ht is defined as follows:

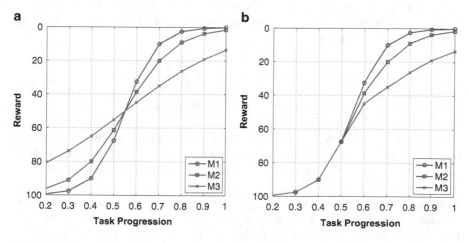

Fig. 5.9 Task rewarding model. (**a**) Initial rewarding model. (**b**) Refined rewarding model

Table 5.2 Rewarding model and related parameters

Symbol	Description
$RW(PT(ht))$	Rewarding function based on the task processing time $PT(ht)$. The output of the function is a percentage value between 0 % and 100 %. 100 % is given if the task is processed very fast. 0 % reward is given is given if the task expires and thus fails
ψ	Saturation of $RW : [0, \tau] \to [0, \psi], \psi \in [0, 1]$
σ	Parameter to define the horizontal displacement of RW
δ	Parameter to define the "steepness" of RW's slope. Steepness means that the reward changes more quickly depending on task processing time
M	Depicts a rewarding model for different parameters

$$RW(PT(ht)) = \frac{\psi}{1 + EXP(-\dfrac{PT(ht) - \sigma}{\delta})} \tag{5.3}$$

A description of the model's parameters is given in Table 5.2:

Figure 5.9a visualizes the output of RW depending on different parameters (Table 5.2). The horizontal axis shows the task progression in terms of processing time. Progression 0 means that the task has not yet started. Progression 1 means that the task has expired (maximum processing time has been reached). Thus, a reward is given between $0 < PT(ht)/MAX(ht) < 1$. The function $MAX(ht)$ returns the maximum processing time of ht.

The basic idea is to use different models (e.g., $M1, M2, M3$) to account for the *risk* that a particular type of task will not be processed in a timely manner. Risk is automatically calculated based on *finished* versus *aborted* tasks within the community (the parameter δ in Table 5.2). The task rewarding function RW should fall less steeply if a particular type of task tends to be aborted by the community. To model risk for the task progression spectrum that is based on the task processing time, RW needs to be refined as a stepwise function RW':

$$RW'(PT(ht)) = \begin{cases} RW(PT(ht)) & , \text{if } \dfrac{PT(ht)}{MAX(ht)} < 0.5 \\ RW(PT(ht); M) & , \text{otherwise} \end{cases} \tag{5.4}$$

Figure 5.9b shows RW'. Progression (based on processing time) towards a particular point (see 0.5 on horizontal axis) results in equal rewards regardless of the model ($M1, M2, M3$). Beyond this point, tasks are differently rewarded depending on the risk modeled by a given model M. For example, given $M3$ that models tasks with higher risks, higher rewards are given because a successfully processed task becomes more valuable for the task creator. The detailed calculation of the risk factor in the model (the parameter δ) is not explained in detail in this work. The detailed mechanism is explained in [33, pp. 85].

The benefit of this approach is that the rewarding function RW undergoes a self-configuration process by selecting a particular model M automatically based on monitored interactions. For example, if many tasks are aborted and the risk of unsuccessful task processing increases, also the reward function RW can be updated and a different model M is selected.

5.6.3   Broker Importance

Here we introduce our broker ranking approach. As mentioned before, the goal of broker importance ranking is to implement the SBQL statement `Order by` (cf. Fig. 5.5). The basic idea of the approach is derived from the concept of hubs and authorities wherein the hub importance of a node in a network is influenced by the authority of the nodes the hub is connected to. This method is also known as the concept Hyperlink-Induced Topic Search (HITS) as introduced by [20]. A node's authority is influenced by the hub importance of the node's neighbors. Let us denote the importance of broker b as $B(b)$ and the importance of community c as $C(c)$. Communities are sets of nodes that are matched and grouped using SBQL. A community/broker graph can be depicted as a graph $G(V, E)$ where V depicts the set of nodes that may be either brokers or communities and E the set of edges to depict links between brokers and communities. Utilizing the idea of hubs and authorities, the broker and community importance is defined as follows:

$$B(b) = \sum_{c \in N(b)} C(c) \qquad C(c) = \sum_{b \in N(c)} B(b) \qquad (5.5)$$

The set of neighbors (either community or brokers neighbors) is depicted by $N(c)$ for the set of brokers that are connected to c and $N(b)$ for the set of communities that are connected to broker b. Hence, b's importance is directly influenced by the importance of the community it is connected to. Similarly, the importance of c is based upon the importance of brokers that are attached to c.

We make some important extensions to the basic model:

- We expand the models for $B(b)$ and $C(c)$ towards a PageRank-like model that is more robust with regards to rank stability to small perturbations (see also [28]) and can be personalized to topics of interest.
- Personalization relates to the previously discussed hierarchical tag clustering since broker or community importance scores can be calculate for certain interest areas (clusters).
- We additionally use broker and community weights. A broker may not equally engage in all communities and thus weights propagate importance scores based on the strength of the broker's community involvement.

The resulting model is depicted by the following two equations (see also Table 5.3):

Table 5.3 Description of symbols

Symbol	Description
$p(b;T)$	The topic-sensitive broker personalization vector for topic T. The parameter λ_b is used to balance between personalization and "network importance"
$p(c;T)$	The topic-sensitive community personalization vector for topic T. Similarly, the parameter λ_c balances between personalization and network importance
w_{bc}^T	The topic-based broker community weight that is based on how much b engages in c (e.g., number of interactions between b and c)

$$B(b;T) = (1 - \lambda_b)p(b;T) + \lambda_b \sum_{c \in N(b)} w_{bc}^T C(c;T) \qquad (5.6)$$

$$C(c;T) = (1 - \lambda_c)p(c;T) + \lambda_c \sum_{b \in N(c)} w_{bc}^T B(b;T) \qquad (5.7)$$

The broker and community scores, $B(b;T)$ and $C(c;T)$ respectively, can be calculated at various levels in the hierarchical tree depending on the topic of interest. However, it is important to note that the topic of interest depends on the actual query and its associated keywords. Therefore, these scores would need to be calculated for every query that is used to discover and rank brokers. For large social and collaborative networks such an approach is not feasible due to computational complexity. For example, for a large network computation of broker scores could talk some days or even weeks (depending on hardware resources). Clearly, these scores need to be computed in an offline manner.

The ultimate goal is that topic-sensitive importance scores are computed offline and at query time aggregated into a composite ranking score. We propose the PageRank linearity theorem to solve the problem of topic-sensitive broker importance ranking. The linearity theorem [16] is defined as:

Theorem 5.1 (Linearity). *For any personalization vectors* p_1, p_2 *and weights* w_1, w_2 *with* $w_1 + w_2 = 1$, *the following equality holds:*

$$PPV(w_1 p_1 + w_2 p_2) = w_1 PPV(p_1) + w_2 PPV(p_2) \qquad (5.8)$$

The above equality states that personalized PageRank vectors PPV can be composed as the weighted sum of PageRank vectors. To utilize the theorem, we need to arrive at a different mathematical representation of the broker importance $B(b;T)$. In its current form the linearity theorem cannot be applied. First, we substitute $C(c;T)$ [see Eq. (5.7)] in $B(b;T)$ [see Eq. (5.6)] so that we have:

$$B(b;T) = (1 - \lambda_b)p(b;T) + \lambda_b(1 - \lambda_c) \sum_{c \in N(b)} w_{bc}^T p(c;T) \qquad (5.9)$$

$$+ \lambda_b \lambda_c \sum_{c \in N(b)} \sum_{b' \in N(c)} w_{bc}^T w_{b'c}^T B(b';Q) \qquad (5.10)$$

Let us define the personalization vector $p'(b;T)$ as follows:

$$p'(b;T) = \frac{(1 - \lambda_b)}{(1 - \lambda_c)} p(b;T) + \lambda_b \sum_{c \in N(b)} w_{bc}^T p(c;T) \tag{5.11}$$

The personalization vector $p'(b;T)$ can be simplified and rewritten if the λ parameters are set to $\lambda_b = \lambda_c$:

$$p'(b;T) = p(b;T) + \lambda \sum_{c \in N(b)} w_{bc}^T p(c;T) \tag{5.12}$$

The personalization vector in Eq. (5.12) has two components: $p(b;T)$ shows the topic specific personalization for broker b and $\sum_{c \in N(b)} w_{bc}^T p(c;T)$ the topic-based personalization of each community the broker b is connected to. Equation (5.13) shows $B(b;T)$ using $p'(b;T)$.

$$B(b;T) = (1 - \lambda)p'(b;T) + \lambda^2 \sum_{c \in N(b)} \sum_{b' \in N(c)} w_{bc}^T w_{b'c}^T B(b';T) \tag{5.13}$$

Equation (5.13) has a PageRank-like structure and thus the linear theorem can be applied. The following Eq. (5.14) shows the query-based aggregation of broker importance scores $B(b;Q)$.

$$B(b;Q) = w_1 B(b;T_1) + w_2 B(b;T_2) \qquad \text{with } Q = \{T_1, T_2\} \tag{5.14}$$

The symbol Q in $B(b;Q)$ depicts the SBQL query that contains the set of demanded topics $T1$ and $T2$, which are used to match communities and brokers. The result of the computation in Eq. (5.14) is a composite broker importance score that is used to rank brokers. Computing composite scores requires only two database read operations to obtain the topic-based scores and aggregation of the respective scores. This greatly reduces the time needed to rank brokers.

5.7 Evaluation

5.7.1 Overview

We performed two kinds of experiments to evaluate SBQL and its broker importance ranking approach.

- Performance tests to evaluate the suitability of SBQL and its implementation to query data obtained from a real service-based testbed environment. The testbed generates instances of services communicating over real WS-stacks and allows for simulation of service behavior (response time, generated faults, etc.) Our evaluations were gathered using the logging features of the Genesis2 framework [17].

- Evaluation of the broker importance ranking approach using real data from a virtual collaboration environment. We used data from the ICT research projects having received grants under the EU's Seventh Framework Programme (FP7) [27]. The data covers a period from 2007 to 2011. Research projects have multiple partners and an organization can be the partner of multiple projects.

These experiments help to evaluate both efficiency of SBQL and quality with respect to broker ranking. First, the performance results are discussed and second the broker ranking results.

5.7.2   Performance Tests

Here we discuss results related to SBQL performance tests obtained by using the Genesis2 service testbed. The setup is described in the following.

- **Testbed environment.** Genesis2 has a management interface and a controllable runtime to deploy, simulate, and evaluate SOA designs and implementations. A collection of extensible elements for these environments are available such as models of services, clients, registries, and other SOA components. Each element can be set up individually with its own behavior, and steered during execution of a test case.
- **Backend deployment.** For the experiments in this work we deployed Genesis2 Backends to the *Amazon Elastic Compute Cloud*. We launched depending on the amount of involved services instances of two or three *Community AMIs* of the type *High-Memory Extra Large Instance* (17.1 GB of memory) running a Linux OS. In the following we provided each instance with the same Genesis2 Backend snapshot via mountable volumes from the *Elastic Block Store*. Finally, we deployed the following environment setup from a local Genesis2 Frontend. It included SOA-based PVCs established by Genesis2 Web services equipped with simulated behavior and predefined relations to provide communication channels and instantiate online communities.

  Services act like HPSs when delegating each other new tasks, processing tasks, re-delegating existing tasks, or reporting tasks' progress status. In other words, by following the HPS concept, services are provided by human actors and thereby exhibit human behavior. Tasks are not delegated arbitrarily but must match the receivers capabilities. Therefore, they are tagged by three keywords one of which must match the picked receivers capabilities. As an intermediate, a broker combines capabilities of the two communities it connects. The broker avoids task processing and only forwards tasks. The finally deployed environments are variable in number of services, number of participants per group (2–5 services) and consequently also in number of communities and required brokers that connect at least each community with another. Task processing and delegation decisions happen individually and in random time intervals (1–8 s).

- **Client setup.** We simulated environments with different numbers of nodes and interactions to obtain insights in performance aspects. SBQL tools and related graph libraries have been implemented in state-of-the-art technology using .NET/C# and have been deployed on our lab-based server infrastructure. The blade servers are equipped with 3.2 GHz quad core CPUs and 10 GB RAM. Interaction logs are managed by MySQL databases. A client request pool is created on a separate machine (Intel Core2 Duo CPU 2.50 GHz, 4 GB RAM) to perform parallel invocations of the SBQL query Web service. Clients are connected with the server via a local 100 MBit Ethernet.

We performed several experiments to test the performance of our SBQL implementation under varying characteristics such as number of nodes and groups. The results are summarized in the following. We performed a set of concurrent queries (50 concurrent queries) time by launching multiple threads. For each load experiment, the total number of requests was 100 requests to be processed. By processing a larger amount of requests, say 200, the total processing time linearly increases with the number of requests.

We increased the number of nodes and interactions to understand the scalability of SBQL under different conditions. The first experiment comprises 198 nodes, 200 edges, and a total number of ten distinct tags applied to interactions between nodes. In experiment 2, we simulated 579 nodes, experiment 3 comprising 774 nodes, and experiment 4 with 1029 nodes in the tested. HPSs in the testbed have been deployed equally on multiple hosts, e.g., three cloud hosts in experiment 4 to achieve scalability. The cloud deployments have been done on Amazon's cloud. The SBQL processing time for this environment is shown in Fig. 5.10.

To compare the experiments 1–4, we query the graph data using the query keywords "Robustness" and "Logging" to obtain a set of matching brokers that are able to broker tasks related to those keywords. Increasing the number of nodes by

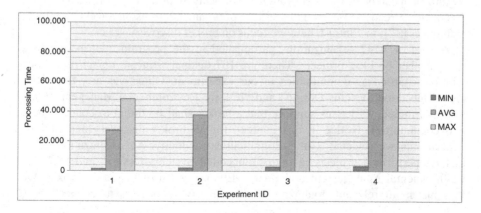

Fig. 5.10 SBQL processing time (in milliseconds)

Table 5.4 SBQL queries using the data from experiment 4, number of discovered brokers and MIN processing time

ID	SBQL query keywords	Num. brokers	MIN time
Q1	"Robustness", "Log"	105	3993
Q2	"Robustness", "Log", "DB", "Testbed"	134	3666
Q3	"Robustness", "Log", "DB", "Testbed", "Similarity"	146	3478

a factor of ≈ 3 (see the experiments 1 and 2), the processing time of SBQL queries increases by 30 %. Comparing the experiments 2 and 3 (node addition of $\approx 30\%$), the processing time increases by a factor of 10 %. By comparing the experiments 3 and 4 (node addition of $\approx 30\%$), the SBQL processing time increases by a factor of 20 %. These experiments show that SBQL scales with larger testbed environments linearly.

To test the effect of using various query keywords, we used different keyword combinations as shown in Table 5.4. The first column shows the query ID, second the SBQL query keywords are shown, third the number of brokers are depicted, and forth the minimum SBQL processing time in milliseconds is shown.

The number of discovered brokers increases if multiple keywords are specified. However, the average SBQL processing time is not significantly influenced by the number of used keywords.

5.7.3   Ranking Experiments

The second set of experiments is based on research projects and partners having received grants under the EU's Seventh Framework Programme (FP7). Detailed statistical information regarding the ICT community is described in [27] and covers a period from 2007 to 2011. Research projects have multiple partners and an organization can be the partner of multiple projects. In prior research, we have used this dataset already to rank the importance of organizations using novel link mining techniques [37]. In contrast, here we focus on broker discovery issues.

A total of 4747 organizations participated in the program contributing to a total number of 1451 projects. There are 107 distinct *strategic objectives*, which will be regarded as communities in our experiment. Objectives are, for example, "Embedded Systems Design" and "ICT for Environmental Management and Energy Efficiency". Let us suppose that a broker shall be found to connect these two communities. Notice in this context, for all broker discovery and ranking experiments we selected 2 to 3 communities randomly. An organization needs to have performed projects in both "Embedded Systems Design" and "ICT for Environmental Management and Energy Efficiency" to qualify as a broker. We first match all relevant organizations that qualify as brokers to connect these

Table 5.5 Matched and top-10 ranked brokers for "ICT for Environmental Management and Energy Efficiency" and "Embedded Systems Design"

Rank	Organization (broker)	PC	Score
1	FRAUNHOFER-GESELLSCHAFT ZUR FOERDERUNG DER ANGEWANDTEN FORSCHUNG E.V	4	0.1928
2	ECOLE POLYTECHNIQUE FEDERALE DE LAUSANNE	6	0.1395
3	POLITECNICO DI MILANO	5	0.1268
4	COMMISSARIAT A L ENERGIE ATOMIQUE ET AUX ENERGIES ALTERNATIVES	6	0.1225
5	UNIVERSITY OF SOUTHAMPTON	2	0.1015
6	INSTITUTE OF COMMUNICATION AND COMPUTER SYSTEMS	2	0.0994
7	UNIVERSITEIT TWENTE	2	0.0889
8	IMPERIAL COLLEGE OF SCIENCE, TECHNOLOGY AND MEDICINE	3	0.0883
9	TECHNISCHE UNIVERSITAET GRAZ	4	0.0771
10	DEUTSCHES ZENTRUM FUER LUFT - UND RAUMFAHRT EV	2	0.0661

communities and then perform ranking using the introduced link based broker importance model [cf. Eq. (5.13)].

The parameter λ is set to $\lambda = 0.85$ (this is the suggested value by the PageRank model [30]), the broker personalization vector is assigned uniformly to $p'(b; T) = \frac{1}{numOrgs}$ where $numOrgs$ depicts the number of organizations (in this case $numOrgs = 4747$), and the weight $w_{bc}^T = \frac{1}{|N(b)|}$ is based on the number of communities $|N(b)|$ (i.e., the number of neighbors in the graph) the broker b is connected to.

The top-10 ranking results are detailed in Table 5.5. The first column shows the rank (1 ... 10), the second column shows the organizations' names, the third column shows the number of projects (project count PC) that the organization has performed in the two objectives ("Embedded Systems Design" or "ICT for Environmental Management and Energy Efficiency"), and the last column shows the numerical broker ranking score. The brokers are sorted according to the ranking score from high to low.

In addition to the description in Table 5.5, Fig. 5.11 visualizes all brokers connecting the two communities. The node size of the top-10 ranked brokers is based on their rank. The organization Fraunhofer-Gesellschaft has in general (within the ICT framework) the largest number of projects and receives the highest amount of funding. Since it is involved in many project, Fraunhofer also has the highest importance because it is connected to many communities. With regards to the topic-sensitive results, it is specifically involved in four projects relevant to the demanded objectives "ICT for Environmental Management and Energy Efficiency" and "Embedded Systems Design". The combination of high reputation and involvement in relevant communities makes Fraunhofer the top-ranked broker. The second ranked broker is Ecole Polytechnique Federale de Lausanne (EPFL)

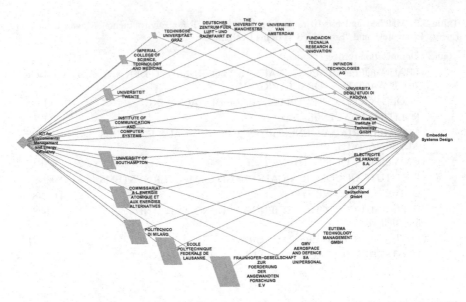

Fig. 5.11 Ranked brokers connecting "ICT for Environmental Management and Energy Efficiency" and "Embedded Systems Design"

Table 5.6 Matched and ranked brokers for "Flexible, Organic and Large Area Electronics and Photonics" and "Embodied Intelligence"

Rank	Organization (broker)	PC	Score
1	CENTRE NATIONAL DE LA RECHERCHE SCIENTIFIQUE	2	0.1398
2	IMPERIAL COLLEGE OF SCIENCE, TECHNOLOGY AND MEDICINE	2	0.0883

with many relevant projects with regards to the objectives but with less "global" importance (within the entire ICT framework) than Fraunhofer. Also, Politecnico di Milano is involved in many relevant projects and thus ranks at position 3.

The broker ranking approach delivers the expected results and combines successfully topic-relevant importance with global network importance.

Another example is shown in Table 5.6 where communities "Flexible, Organic and Large Area Electronics and Photonics" and "Embodied Intelligence" can be reached via two brokers. Both have two projects (one in each objective) and thus qualify as brokers.

A final example is shown by Table 5.7 where brokers need to connect three communities. Indeed, the number of discovered brokers varies depending on community (objective) popularity. To conclude the discussion on ranking results, the proposed broker importance model delivers expected results by combining global network importance of organizations with topic-based relevance. Thus, our approach is able to (a) identify brokers that formally match the search criteria and (b) rank brokers according to importance.

Table 5.7 Matched and ranked brokers for "ICT for Environmental Management and Energy Efficiency" and "Intelligent information management" and "Micro/nanosystems"

Rank	Organization (broker)	PC	Score
1	FRAUNHOFER-GESELLSCHAFT ZUR FOERDERUNG DER ANGEWANDTEN FORSCHUNG E.V	14	0.1928
2	ECOLE POLYTECHNIQUE FEDERALE DE LAUSANNE	9	0.1395
3	POLITECNICO DI MILANO	3	0.1268
4	UNIVERSIDAD POLITECNICA DE MADRID	5	0.1186
5	UNIVERSITY OF SOUTHAMPTON	5	0.1015
6	STIFTELSEN SINTEF	3	0.0813

5.7.4 Lessons Learned

This section provides some additional discussions with regards to lessons learned and recommendations for theory and practice.

- SBQL provides a rich set of language features to state complex queries for broker discovery. Distributed queries need to be considered also in future SBQL activities. Currently it is assumed that logs are collected in a central repository and that SBQL queries are executed on a single instance database. In this regard, distributed log repositories and databases should be considered in SBQL queries.
- The performance tests showed that our SBQL implementation offers sufficient performance for mid- to large-scale environments. In truly open ultra large-scale environments with potentially millions of people and services performance and scalability need to be revisited. Aspects include distribution of queries and federation of results.
- The broker importance ranking approach delivers excellent results and allows for full personalization. The mathematical models and theories, as presented in this work, provide mature concepts and the basis for further work on personalization techniques.
- Further personalization may include the long-term stability of social relations, for example, or the frequency of changes of collaboration partners. The weight assignment of social relations is an important issue and should be further analyzed by considering various VO datasets.

5.8 Conclusions

This work introduced a number of principles to support service oriented collaboration in professional virtual communities. The problem in today's collaborative networks is the fragmented nature of expertise areas and boundaries imposed by organizational structures. Principles found in social network theory are suitable for assisting in the formation process of socially-based compositions of human and software services. Specifically, we adopted the well-established theory of structural holes to support the formation of such compositions. In this work, brokers help connecting independent communities by forwarding tasks and help and support requests to individual community members. Here we proposed a socially-based approach to discover brokers based on interaction mining and monitoring techniques. Technically, we proposed SBQL to discover suitable brokers based on query constraints and ranking criteria. SBQL introduces a domain specific language to construct and execute complex queries specifically for the discovery of brokers in socially-based virtual communities. Here we introduced a broker importance model not only to match brokers but also to rank them according to their topic-based and community-wide relevance. Periodically updated metrics are used to weight interaction links and paths between actors. The proposed techniques and technical concepts have been implemented and validated through a services testbed and through experiments using real world data.

In our future work we will further work on tool support for broker query modeling and debugging. Furthermore, we plan to use further additional social network data sources to perform broker ranking experiments. Also, we plan to support the execution of distributed SBQL queries.

References

1. Gautam Ahuja. Collaboration networks, structural holes, and innovation: A longitudinal study. *Administrative Science Quarterly*, 45(3):425–455, September 2000. ISSN 00018392. doi: 10.2307/2667105.
2. Donovan Artz and Yolanda Gil. A survey of trust in computer science and the semantic web. *J. Web Sem.*, 5(2):58–71, 2007.
3. Ashish Agrawal et al. WS-BPEL Extension for People (BPEL4People)., 2007.
4. Lars Backstrom, Dan Huttenlocher, Jon Kleinberg, and Xiangyang Lan. Group formation in large social networks: membership, growth, and evolution. In *12th ACM SIGKDD conference on knowledge discovery and data mining*, KDD '06, pages 44–54. ACM, 2006.
5. Marco Bertoni, Andreas Larsson, Åsa Ericson, Koteshwar Chirumalla, Tobias Larsson, Ola Isaksson, and Dave Randall. The rise of social product development. *IJNVO*, 11(2):188–207, 2012.
6. Stephen P. Borgatti, Ajay Mehra, Daniel J. Brass, and Giuseppe Labianca. Network analysis in the social sciences. *Science*, 323(5916):892–895, 2009. doi: 10.1126/science.1165821.
7. Ronald S. Burt. Structural holes and good ideas. *American Journal of Sociology*, 110(2): 349–399, September 2004.
8. Ronald S. Burt. *Brokerage and Closure: An Introduction to Social Capital*. Oxford University Press, USA, October 2005.
9. Luis M. Camarinha-Matos and Hamideh Afsarmanesh. Collaborative networks. In *PROLA-MAT*, pages 26–40, 2006.
10. Catherine Dwyer, Starr Roxanne Hiltz, and Katia Passerini. Trust and privacy concern within social networking sites: A comparison of facebook and myspace. In *AMCIS*, 2007.
11. Jennifer Golbeck. Trust and nuanced profile similarity in online social networks. *TWEB*, 3(4), 2009.
12. Tyrone Grandison and Morris Sloman. A survey of trust in internet applications. *IEEE Communications Surveys and Tutorials*, 3(4), 2000.

13. R. Guha, Ravi Kumar, Prabhakar Raghavan, and Andrew Tomkins. Propagation of trust and distrust. In *WWW*, pages 403–412, 2004.
14. Roger Guimera, Brian Uzzi, Jarrett Spiro, and LuÃÑs A. Nunes Amaral. Team assembly mechanisms determine collaboration network structure and team performance. *Science*, 308 (5722):697–702, 2005. doi: 10.1126/science.1106340.
15. Ralf Hartmut Güting. Graphdb: Modeling and querying graphs in databases. In *Proceedings of the 20th International Conference on Very Large Data Bases*, VLDB '94, pages 297–308, San Francisco, CA, USA, 1994. Morgan Kaufmann Publishers Inc.
16. Taher H. Haveliwala. Topic-sensitive pagerank. In *WWW*, pages 517–526, 2002.
17. Lukasz Juszczyk and Schahram Dustdar. Script-based generation of dynamic testbeds for soa. In *International Conference on Web Services*, 2010.
18. Eva C. Kasper-Fuehrera and Neal M. Ashkanasy. Communicating trustworthiness and building trust in interorganizational virtual organizations. *Journal of Management*, 27(3):235–254, June 2001. doi: 10.1177/014920630102700302.
19. Florian Kerschbaum, Jochen Haller, Yücel Karabulut, and Philip Robinson. Pathtrust: A trust-based reputation service for virtual organization formation. In *iTrust*, pages 193–205, 2006.
20. Jon Kleinberg. Authoritative sources in a hyperlinked environment. *Journal of the ACM*, 46: 668–677, 1999.
21. Jon Kleinberg, Siddharth Suri, Éva Tardos, and Tom Wexler. Strategic network formation with structural holes. *ACM Conference on Electronic Commerce*, 7(3):1–4, 2008.
22. Xin Li, Lei Guo, and Yihong Eric Zhao. Tag-based social interest discovery. In *WWW '08: Proceeding of the 17th international conference on World Wide Web*, pages 675–684, New York, NY, USA, 2008. ACM.
23. Tiancheng Lou and Jie Tang. Mining structural hole spanners through information diffusion in social networks. In *Proceedings of the 22nd international conference on World Wide Web*, WWW '13, pages 825–836, 2013.
24. Miriam J. Metzger. Privacy, trust, and disclosure: Exploring barriers to electronic commerce. *J. Computer-Mediated Communication*, 9(4), 2004.
25. Thomas P. Moran, Alex Cozzi, and Stephen P. Farrell. Unified activity management: Supporting people in e-business. *Com. of the ACM*, 48(12):67–70, 2005.
26. Lik Mui, Mojdeh Mohtashemi, and Ari Halberstadt. A computational model of trust and reputation for e-businesses. In *HICSS*, page 188, 2002.
27. Filippo Munisteri. ICT statistical report for annual monitoring 2011. http://ec.europa.eu/digital-agenda/sites/digital-agenda/files/stream_2012_0.pdf, February 2012.
28. Andrew Y. Ng, Alice X. Zheng, and Michael I. Jordan. Stable algorithms for link analysis. In *Proceedings of the 24th annual international ACM SIGIR conference on Research and development in information retrieval*, SIGIR '01, pages 258–266, New York, NY, USA, 2001. ACM.
29. Bart Nooteboom, Victor Gilsing, Wim Vanhaverbeke, Geert Duijsters, and Ad Oord. Network embeddedness and the exploration of novel technologies: Technological distance, betweenness centrality and density. Technical Report 2006-32, Tilburg University, Center for Economic Research, 2006.
30. Lawrence Page, Sergey Brin, Rajeev Motwani, and Terry Winograd. The pagerank citation ranking: Bringing order to the web. Technical report, Stanford University, 1998.
31. Joel M. Podolny et al. Resources and relationships: Social networks and mobility in the workplace. *American Sociological Review*, 62:673–693, 1997.
32. Royi Ronen and Oded Shmueli. Soql: A language for querying and creating data in social networks. In *ICDE*, pages 1595–1602, 2009.
33. Daniel Schall. *Human Interactions in Mixed Systems - Architecture, Protocols, and Algorithms*. PhD thesis, Vienna University of Technology, 2009.
34. Daniel Schall. A human-centric runtime framework for mixed service-oriented systems. *Distributed and Parallel Databases*, 29(5-6):333–360, 2011.
35. Daniel Schall. Expertise ranking using activity and contextual link measures. *Data Knowl. Eng.*, 71(1):92–113, 2012a. doi: 10.1016/j.datak.2011.08.001.

36. Daniel Schall. *Service Oriented Crowdsourcing: Architecture, Protocols and Algorithms.* Springer Briefs in Computer Science. Springer New York, New York, NY, USA, 2012b. doi: 10.1007/978-1-4614-5956-9.
37. Daniel Schall. Measuring contextual partner importance in scientific collaboration networks. *Journal of Informetrics*, 7(3):730–736, 2013a. doi: 10.1016/j.joi.2013.05.003.
38. Daniel Schall. Formation and interaction patterns in social crowdsourcing environments. *Int. J. Communication Networks and Distributed Systems*, 11(1):42–58, 2013b.
39. Daniel Schall, Hong-Linh Truong, and Schahram Dustdar. Unifying human and software services in web-scale collaborations. *IEEE Internet Computing*, 12(3):62–68, 2008.
40. Daniel Schall, Florian Skopik, Harald Psaier, and Schahram Dustdar. Bridging socially-enhanced virtual communities. In *ACM SAC*, pages 792–799, 2011.
41. Diego Serrano, Eleni Stroulia, Denilson Barbosa, and Victor Guana. Sociql: A query language for the socialweb. In Evangelos Kranakis, editor, *Advances in Network Analysis and its Applications*, volume 18 of *Mathematics in Industry*, pages 381–406. Springer Berlin Heidelberg, 2013.
42. Ahm Shamsuzzoha, Timo Kankaanpaa, Luis Maia Carneiro, and Ricardo Almeida. Methodological support to establish a collaborative non-hierarchical business network for complex product manufacturing. *IJNVO*, 11(3/4):363–381, 2012.
43. Andriy Shepitsen, Jonathan Gemmell, Bamshad Mobasher, and Robin Burke. Personalized recommendation in social tagging systems using hierarchical clustering. In *RecSys '08: Proceedings of the 2008 ACM conference on Recommender systems*, pages 259–266, New York, NY, USA, 2008. ACM.
44. Florian Skopik, Daniel Schall, and Schahram Dustdar. Modeling and mining of dynamic trust in complex service-oriented systems. *Information Systems*, pages 735–757, 2010. doi: doi:10.1016/j.is.2010.03.001.
45. W3C. Sparql query language for rdf, 2008.
46. Duncan J. Watts and Steven H. Strogatz. Collective dynamics of small-world networks. *Nature*, 393(6684):409–10, 1998.
47. Brian Whitworth and Cheickna Sylla. A social environmental model of socio-technical performance. *IJNVO*, 11(1):1–29, 2012.
48. Cai-Nicolas Ziegler and Jennifer Golbeck. Investigating interactions of trust and interest similarity. *Decision Support Systems*, 43(2):460–475, 2007.

Chapter 6
Conclusion

Online social networks have become an integral part of our daily personal and work related activities. People use social networks not only to maintain relationships with friends but also to communicate, collaborate, and share information. One of the most profound properties of social networks is their dynamic nature due to people joining and leaving networks. This book introduced novel techniques for link formation in social network based systems.

First, this book introduced link prediction in directed social networks. The prediction of missing links and the prediction of future links is an important task in the domain of social network analysis. The former helps to infer the "real" social network structure while the latter is used to give friendship as well as following recommendations to users. A wide range of local, global, and semi-local metrics have been proposed by previous work. A large body of existing literature, however, focuses on undirected networks only. This work closes this gap by focusing on directed networks. The approach is based on local network information wherein prediction is performed based on node similarity.

The following two chapters introduced link formation through recommendation based on global network information. Recommendation is based on the concept of authority and also structural holes. We proposed a novel follow recommendation approach that is based on the concept of user authority. Instead of simply matching users by static skill profiles, we proposed a network-centric approach taking a user's community engagement as well as social metrics into account. We have systematically derived a mathematically sound model to measure user authority based on activity (e.g., repository commits) and community reputation (follower degree). Furthermore, this work introduced various metrics for importance ranking in scientific collaboration environments. We proposed a novel topic-sensitive authority model that is based on well-establish ranking techniques. We systematically derived a unified HITS/PageRank-based model that can be fully personalized. The second metric measures organizations' structural importance based on the notion of structural holes. In our approach structural importance is computed with

© Springer International Publishing Switzerland 2015
D. Schall, *Social Network-Based Recommender Systems*,
DOI 10.1007/978-3-319-22735-1_6

respect to certain topics of interest. Thus, structural importance helps identifying organizations that may be valuable partners for strategic alliances. Combined with authority, this provides a powerful approach for ranking and discovering new partners. Finally, authority and structural importance are systematically combined with cost. For that purpose we utilize AHP to achieve a trade-off among various ranking criteria. The proposed approach delivers very good results and provides more accurate, topic-sensitive results when compared with other ranking techniques.

The notion of brokers is a hybrid (semi-local) formation approach wherein local information is used to constraint which nodes may act as brokers and global information to rank brokers based on their reputation. Technically, we proposed SBQL to discover suitable brokers based on query constraints and ranking criteria. SBQL introduces a domain specific language to construct and execute complex queries specifically for the discovery of brokers in socially-based virtual communities. Here we introduced a broker importance model not only to match brokers but also to rank them according to their topic-based and community-wide relevance. Periodically updated metrics are used to weight interaction links and paths between actors. The proposed techniques and technical concepts have been implemented and validated through a services testbed and through experiments using real world data.

Future work will focus more on deep learning techniques including adaptive random forest models that can be incrementally improved.

Printed in the United States
By Bookmasters